Nuclear Power

Recent Titles in the
CONTEMPORARY WORLD ISSUES
Series

Books in the **Contemporary World Issues** series address vital issues in today's society such as genetic engineering, pollution, and biodiversity. Written by professional writers, scholars, and nonacademic experts, these books are authoritative, clearly written, up-to-date, and objective. They provide a good starting point for research by high school and college students, scholars, and general readers as well as by legislators, businesspeople, activists, and others.

Each book, carefully organized and easy to use, contains an overview of the subject, a detailed chronology, biographical sketches, facts and data and/or documents and other primary source material, a directory of organizations and agencies, annotated lists of print and nonprint resources, and an index.

Readers of books in the **Contemporary World Issues** series will find the information they need in order to have a better understanding of the social, political, environmental, and economic issues facing the world today.

CONTEMPORARY WORLD ISSUES

Science, Technology, and Medicine

Nuclear Power

A REFERENCE HANDBOOK,
SECOND EDITION

Harry Henderson

ABC-CLIO

Santa Barbara, California • Denver, Colorado • Oxford, England

I would like to thank my ABC-CLIO editors Robin Tutt and David Paige for their never failing and patient assistance in every aspect of writing this book, but especially in obtaining outside contributions for the "Perspectives" chapter.

Library of Congress Cataloging-in-Publication Data

Henderson, Harry, 1951–
   Nuclear power : a reference handbook / Harry Henderson. — Second edition.
      p. cm — (Contemporary world issues)
   Includes bibliographical references and index.
   ISBN 978–1–61069–396–7 (hardcover : acid-free paper) — ISBN 978–1–61069–397–4 (ebook) 1. Nuclear engineering—Handbooks, manuals, etc. 2. Nuclear engineering—History—Handbooks, manuals, etc. 3. Nuclear accidents—Handbooks, manuals, etc. 4. Nuclear engery—Government policy—Handbooks, manuals, etc. I. Title.
TK9146.H45   2014
333.792′409—dc23   2014005270

ISBN: 978–1–61069–396–7
EISBN: 978–1–61069–397–4

18 17 16 15 14   1 2 3 4 5

This book is also available on the World Wide Web as an eBook.
Visit www.abc-clio.com for details.

ABC-CLIO, LLC
130 Cremona Drive, P.O. Box 1911
Santa Barbara, California 93116-1911

This book is printed on acid-free paper ∞

Manufactured in the United States of America

For more than half a century, nuclear power has been the subject of promises based on its potential and of environmental concerns and considerations of safety, risk, and economic viability. As an industry that came of age in the second half of the twentieth century, nuclear energy has often been intertwined with Cold War politics, the ambitions of nations seeking a place among the world powers, and even the aims of terrorists and shadowy arms traders.

The history of nuclear energy has been punctuated by a few familiar names that recall times and places where things went very wrong—Three Mile Island, Chernobyl, and most recently, the meltdowns following the 2011 earthquake and tsunami in eastern Japan. A loosely organized but often effective international community of antinuclear activists have fought (sometimes successfully) to shut down nuclear plants or to prevent them from being built in the first place. As of 2013 the United States, major European industrial countries such as France and Germany, as well as China and of course Japan have all been re-evaluating their commitment to nuclear power. They are not reaching the same conclusions, and even within a given country, nuclear policy often changes according to political change and changing economic needs.

At the same time nuclear power advocates have made arguments that are hard to dismiss. In the 1970s and 1980s the hope was raised that nuclear power would reduce or eliminate America's dependence on supplies of oil from countries (particularly in the Middle East) that were subject to political instability and anti-American sentiments—not to mention the possible economic pressure from an oil cartel.

Today new technologies are making oil and gas more accessible from domestic or neighboring sources, but in the 1990s a new concern arose—what to do about the increasingly recognized process of climate change brought about by the carbon dioxide and other greenhouse gases produced by conventional fossil fuel power plants. Now the debate focuses on whether nuclear power might still be the only practicable way to wean the economy from fossil fuels and whether it can be made safe enough, whether nuclear materials can be secure enough from proliferation and terrorism, and whether the growing pools and piles of nuclear waste can be safely stored for hundreds of years to come.

This edition of *Nuclear Power* has been thoroughly revised and considerably expanded to reflect the developments in more than a decade since the publication of the first edition in 2000. The first chapter gives an overview of the history of the development of nuclear power from its roots in physicists' discoveries to the Manhattan Project that created the atomic bomb, "Atoms for Peace," nuclear diplomacy and the Cold War, and accidents from Browns Ferry to Fukushima and how they have shaped the nuclear debate.

Chapter 2 surveys the economic and technical challenges faced by nuclear power and the controversies surrounding radiation effects, safety risks, managing nuclear waste and the fuel cycle, new reactor designs and their possible benefits, and nuclear power as a solution to climate change.

Chapter 3 provides essays from a variety of viewpoints that probe more deeply into today's nuclear debate. It is followed by Chapter 4, which consists of brief profiles of individuals and organizations that have played important roles in the history of nuclear power and the present-day debate.

Chapter 5 presents a wide-ranging collection of primary source documents for nuclear policy, as well as facts and statistics on the nuclear industry; basic concepts of radiation and nuclear fission; illustrations of reactor designs, the fuel cycle, and waste storage; the basic structure of nuclear regulation; and statements from nuclear advocates and critics debating the future of nuclear power.

Chapter 6 is an extensive bibliography of both print and on-line resources, divided into nine topics. Chapters 7 and 8 present a chronology of events and a glossary of important nuclear terms.

No work, and certainly no single work, can touch on every aspect of such a complicated subject, nor can it anticipate all events in a rapidly changing world. However, the background, history, concepts, data, and viewpoints offered here will enable readers to understand and confidently research the ongoing chapters in the history of nuclear power.

Nuclear Power

Born during World War II out of a secret search for a weapon of unprecedented power, the seeds of nuclear energy were planted through a series of remarkable discoveries by physicists more than 100 years ago.

Atoms and Mysterious Rays

What is nuclear energy, and how did people learn how to use it? Most of the energy we use today comes from the chemical process of combustion (burning oil, coal, or natural gas). By the late nineteenth century scientists understood how combustion and other basic chemical reactions work. Atomic theory, proposed by John Dalton at the start of the nineteenth century, explained that all matter is made of atoms of a number of basic substances, called elements. Each element has a particular weight and other characteristics, and can combine with other elements in various ways to form compounds. The periodic table, developed by Dmitri Mendeleev in the 1860s and later refined, showed remarkable patterns in how groups of elements behaved. There was no satisfactory physical explanation for this behavior because, at the time, atoms themselves were believed

Nobel prize-winning physicist Enrico Fermi played a key role in turning particle theory into a new field—nuclear engineering. (Library of Congress)

to be indivisible (which is what the word "atom" means in Greek).

In the mid-1890s, however, a series of startling discoveries began to radically change our understanding of the nature of matter. In 1895 a German physicist, Wilhelm Roentgen, was experimenting with a Crookes (cathode ray) tube like those that would later be used in television sets. He accidentally discovered a strange form of invisible energy that could pass through solid objects. He called it X-rays, with the "X" standing for "unknown."

X-rays for the first time gave health care providers a way to see inside the human body without cutting it open. By World War I, X-rays were revolutionizing the treatment of broken bones and other injuries. But it also turned out that this form of radiation was a powerful tool that allowed physicists to probe the structure of matter.

In 1896 French physicist Antoine Becquerel was wondering whether X-rays might be emitted by substances that had absorbed sunlight. He chose to investigate an interesting substance called uranium, which had been identified by German chemist Martin Heinrich Klaproth in 1789. This gray metal, the heaviest of the natural elements, had greatly intrigued Klaproth, so much that he declared, "I am certain that the investigation of uranium, starting from its natural resources, will lead to many new discoveries. I confidently recommend that one who is looking for new study topics to give particular consideration to the uranium compounds" (Inam ur Rehman 1993, 77).

A century later, Becquerel exposed a compound containing uranium to the sun then placed it on a photographic plate that had been covered with black paper. Although light could not reach the film, it was exposed anyway. Becquerel at first assumed that the uranium, after absorbing sunlight, had emitted X-rays that passed through the paper to the film. But much to his surprise, the uranium exposed the film even after a series of cloudy, sunless days. The uranium itself was producing the energy, but how?

The scientific husband-and-wife team of Marie and Pierre Curie painstakingly sifted through uranium ores in search of the mysterious energy that became known as radioactivity. They found that the uranium itself could not account for all the radiation, and they eventually found new sources of radiation: the elements thorium, polonium, and radium. Marie Curie soon realized that "the radioactivity of these substances [is] decidedly an atomic property. It seems to depend on the presence of atoms of the ... elements in question, and is not influenced by any change of physical state or chemical decomposition" (Grady 1992, 34). In fact, she had discovered a new kind of energy: atomic energy.

The Atomic Nucleus

But weren't atoms featureless, indivisible balls of primal stuff? How could they spontaneously generate energy? It soon became clear that the interior of the atom was actually a complex world containing smaller particles. In 1897 British physicist Joseph John Thompson discovered a negatively charged particle called the electron that was smaller than an atom. Indeed, the electron turned out to be the part of the atom that defined its chemical characteristics (largely explaining the periodic table) and that was responsible for the flow of energy we call electricity.

But what was at the core, or nucleus, of the atom? A New Zealand–born British physicist named Ernest Rutherford, who started his career as Thompson's assistant in the burgeoning new field of X-ray physics, soon found out. Carefully measuring the radiation coming from uranium, Rutherford found that his observations showed "that the uranium radiation is complex, and that there are at present at least two distinct types of radiation—one that is very readily absorbed, and the other of a more penetrating character" (De Andrade 1964).

Rutherford called the positively charged particles alpha, and those that were negatively charged were called beta particles—basically, energetic, free-moving electrons. (In 1900 French physicist Paul Villard discovered a kind of radiation that was not a particle but a form of energy like light but with a much shorter wavelength. He called it gamma rays.)

In 1919 Rutherford and his associate Frederick Soddy began to bombard atoms with fast-moving alpha particles. When an alpha particle hit a nitrogen atom, for example, it was sometimes absorbed. The nitrogen atom instantly emitted a positively charged particle that was equivalent to the nucleus of hydrogen, the simplest of the elements. This particle became known as the proton, for the Greek word for "first." Meanwhile, the nitrogen, retaining one proton, had changed to a heavier element, oxygen. Scientists had discovered that atoms could actually be transmuted, or changed from one element to another, by a change in the number of protons. During the 1920s Rutherford and other researchers discovered other nuclear reactions. Rutherford's measurements also made him realize that "the energy latent in the atom must be enormous compared to that rendered free in ordinary chemical change" (De Andrade 1964, 72).

Even before World War I pioneer science fiction writer H. G. Wells began to explore the tremendous potential of nuclear energy. In his novel *The World Set Free*, he even used the term "atomic bomb" and suggested that people and nations would be unable to restrain the use of their destructive power, even to prevent the destruction of civilization itself:

> Certainly it seems now that nothing could have been more obvious to the people of the earlier twentieth century than the rapidity with which war was becoming impossible. And as certainly they did not see it. They did not see it until the atomic bombs burst in their fumbling hands ... All through the nineteenth and twentieth centuries the amount of energy that men were able to command was continually increasing. Applied to warfare that meant that the power to inflict a blow, the power to

destroy, was continually increasing ... There was no increase whatever in the ability to escape ... Destruction was becoming so facile that any little body of malcontents could use it. (Wells 1914, 103–104)

Fortunately for subsequent generations, the nuclear world war Wells describes has not happened, nor, yet, at least, has there been something akin to the "suitcase bomb" used by terrorists. Nevertheless, the overall picture of increased destructiveness of weapons and the struggle to control their proliferation was remarkably prescient.

Controlled nuclear power, as opposed to an explosive bomb, would also start to appear in science fiction in the 1930s. But to turn atoms into an energy source, there had to be a way to start and control a steady series of nuclear reactions.

Neutrons and Nuclear Fission

The key to controlled nuclear reactions was the neutron, discovered by British physicist James Chadwick in 1932. Physicist Isidor I. Rabi pointed out the neutron's potential use for creating nuclear reactions:

Since the neutron carries no charge, there is no strong electrical repulsion to prevent its entry into nuclei. In fact, the forces of attraction which hold nuclei together may pull the neutron into the nucleus. When a neutron enters a nucleus, the effects are about as catastrophic as if the moon struck the earth. The nucleus is violently shaken up by the blow, especially if the collision results in the capture of the neutron. A large increase in energy occurs and must be dissipated, and this may happen in a variety of ways, all of them interesting. (quoted in Rhodes 1986, 209)

Neutrons turned out to be great "bullets" for bombarding heavy atoms to see what might happen. In 1934 Italian physicist Enrico Fermi bombarded uranium with neutrons, resulting in the creation of new radioactive isotopes of elements near uranium on the periodic table. (An isotope of an element is a variety that has a different number of neutrons in the nucleus. For example, each uranium atom has 92 protons, but different uranium atoms can have different numbers of neutrons, giving atomic weights of 234, 235, or 238. Different isotopes of an element behave the same chemically but can have different effects in nuclear reactions.)

Physicists expected that neutron bombardment would chip away at heavy nuclei, yielding nearby elements whose mass was almost as great as that of the original atom. However, in France, Frederic and Irene Joliot-Curie (the next generation of the remarkable Curie family) had done neutron experiments that seemed to be detecting smaller (medium weight) atoms. Meanwhile, German physicist Otto Hahn and his assistant Fritz Strassman had been bombarding uranium atoms with neutrons. At first, they thought they were getting radium (not unexpected, since radium is only a bit lighter than uranium). However, when Hahn's team tried to verify this, they found they had only barium, a much lighter element that is chemically similar to radium.

Puzzled at these results, Hahn wrote to a colleague, Austrian physicist Lise Meitner (along with Marie Curie, just about the only prominent woman in the field at the time). Meitner replied:

> Your radium results are very startling. A reaction with slow neutrons that supposedly leads to barium! At the moment the assumption of such a thoroughgoing breakup seems very difficult to me, but in nuclear physics we have experienced so many surprises, that one cannot unconditionally say: it is impossible. (quoted in Sime 1997, 235)

A bit later, Meitner and her nephew Otto Frisch discussed Hahn's letter while walking on the snow during a winter's day in 1938. They thought of an earlier idea by the great theoretician Nils Bohr, who had suggested that atomic nuclei might actually split apart:

> We walked up and down in the snow ... and gradually the idea took shape that this was no chipping or cracking of the nucleus but rather a process to be explained by Bohr's idea that the nucleus is like a liquid drop; such a drop might elongate and divide itself ... (quoted in Sime 1997, 236)

Meitner and Frisch soon realized that some uranium atoms were not being merely chipped, but split in two. They used the biological term "fission" to describe this process.

A Letter of Warning

Edward Teller, responding to the spread of the news about nuclear fission, cautioned:

> I believe that urgent action [to maintain secrecy] is required. Very many people have discovered already what is involved ... I repeat there is a chain-reaction mood in Washington. I only had to say "uranium" and then could listen for two hours to their thoughts ... (quoted in Rhodes 1986, 290)

As the news about nuclear fission spread, its implications were quickly grasped by the worldwide community of physicists. Nuclear fission changed a small amount of the mass in the target atom into energy, and Albert Einstein's famous equation $E = MC^2$ indicated that the energy released by even that small mass would be tremendous. Further, each fission was also accompanied by the emission of one or more neutrons. If uranium atoms could be arranged so that the neutrons emitted by each fission triggered another fission, the result could be a controlled chain reaction that could produce abundant, cheap energy. But if the atoms were arranged so that each fission's neutrons produced on the average more than one additional fission, the result would be an uncontrolled chain reaction—and an explosion far greater than that achievable by any chemical explosive. Suppose such a weapon fell into the hands of German chancellor Adolf Hitler, who was even then preparing to launch World War II?

Leo Szilard, Edward Teller, and Eugene Wigner were three Hungarian physicists who had come to the United States at least in part to escape the growing menace of Hitler's Germany. They decided they needed to get the attention of the U.S. government. They were not that prominent in the American consciousness, but fortunately, they knew someone who was—Albert Einstein. Einstein was also a European refugee and had settled to work at the Institute for Advanced Study at Princeton University. Einstein was also the closest thing to a scientific "superstar," his name instantly recognizable to just about any American who had even thought about science.

The physicists prepared a letter to President Franklin D. Roosevelt (it was largely composed by Szilard). The letter warned:

> In the course of the last four months it has been made probable—through the work of Joliot in France as well as Fermi and Szilard in America—that it may become possible to set up a nuclear chain reaction in a large mass of uranium, by which vast amounts of power and large quantities of new radium-like elements would be generated. Now it appears almost certain that this could be achieved in the immediate future. This new phenomenon would also lead to the construction of bombs, and it is conceivable—though much less certain—that extremely powerful bombs of a new type may thus be constructed. A single bomb of this type, carried by boat and exploded in a port,

might very well destroy the whole port together with some of the surrounding territory. However, such bombs might very well prove to be too heavy for transportation by air. (Quoted from Einstein-FDR correspondence at FDR Library, Marist University, http://docs.fdrlibrary.marist .edu/psf/box5/a64a01.html)

If that were not alarming enough, the letter went on to note:

I understand that Germany has actually stopped the sale of uranium from the Czechoslovakian mines which she has taken over. That she should have taken such early action might perhaps be understood on the ground that the son of the German Under-Secretary of State, von Weizsäcker, is attached to the Kaiser-Wilhelm-Institut in Berlin where some of the American work on uranium is now being repeated. (Einstein 1939)

The Manhattan Project

The scientists' letter did get Roosevelt's attention. The American president ordered that the Advisory Committee on Uranium be established to confer further with Szilard, Wigner, and Teller and explore the plausibility of a uranium bomb. In November 1939 the committee reported to Roosevelt that they believed uranium could indeed be used to "provide a possible source of bombs with a destructiveness vastly greater than anything now known" (Hewlett 1962).

Lyman Briggs of the National Bureau of Standards and head of the uranium committee then proposed that the National Defense Research Committee (NRDC) spend $167,000 on research into nuclear fission. They would focus on the two most likely materials, the uranium-235 isotope and the recently discovered element plutonium. The effort would be under the umbrella of the newly created Office of Scientific Research and Development. The rapidity of these developments suggests not only that nuclear fission was being taken seriously, but that a number of American scientists and engineers were becoming convinced that in the increasingly likely next world war, science, technology, and engineering would play a prominent part.

Meanwhile, the science was progressing. By the end of 1939 they knew that about 10 kilograms (22 pounds) of uranium-235 would be enough to enable a nuclear chain reaction. This meant that if they could make an atomic bomb, it would be small enough to carry in and drop from a bomber plane. (British researchers were coming to similar conclusions, and their small program would soon be folded into the American effort.)

In October 1941—about two years after the war had started in Europe—Roosevelt approve a program to develop an atomic bomb. Security, of course, would be an important priority. They had already dropped the word "uranium" from the names of committees and programs. The full-fledged bomb program would be referred to as the Manhattan Engineering District, better known to history as the Manhattan Project.

The scale of the project was unprecedented. It would come to employ more than 130,000 individuals and cost nearly $2 billion (about $26 billion in 2013 dollars). The bulk of this money and effort would go to producing and refining the fissionable materials needed for the bombs. Research, production, and other activities would take place at more than 30 sites— mostly in the United States but also in Canada and the United Kingdom. The biggest sites for fuel production would be at Oak Ridge, Tennessee, and Hanford, Washington.

Historian Richard Rhodes notes, "The Manhattan District bore no relation to the industrial or social life of our country; it was a separate state, with its own airplanes and its own factories and its thousands of secrets" (Rhodes 1986, 277). It is even more remarkable to consider that this massive effort took place simultaneously with fighting World War II in the Pacific and in Europe, including the mounting of the D-day invasion of Normandy.

The First Nuclear Reactor

To build the bomb, scientists and engineers had to solve all the fundamental problems of nuclear power. Most fundamentally, they needed to produce mass quantities of fissionable fuel. They developed techniques to separate the fissionable uranium-235 isotope from the much more abundant uranium-238 by filtering it from the gas uranium hexafluoride. Because they weren't sure if they could get enough fissionable material that way, a separate effort sought to manufacture the fissionable artificial element plutonium-239. That effort required the development of a controlled fission reaction in a nuclear reactor.

The first device to achieve a controlled nuclear chain reaction did not look much like today's nuclear power plants. Built under the stands of the Alonzo Stagg Field stadium at the University of Chicago, it was called the Chicago Pile (It was indeed a "pile" of uranium and graphite blocks—a "moderator" for slowing down neutrons to encourage further fissions; the term "reactor" was not used until the early 1950s). Along with Leo Szilard, its principle developer was Enrico Fermi. Like so many top physicists, Fermi had fled to America to escape fascism and racial laws (in this case Italy's). Fermi was perhaps the most comprehensive of the nuclear physicists in the United States, having made contributions to statistical mechanics and quantum theory as well as nuclear reactions.

All together the pile contained more than 6 tons of pure uranium, 34 tons of uranium oxide, and almost 400 tons of black graphite bricks. The basic design principle was to arrange the fissionable (or "fissile") uranium and the neutron-absorbing graphite blocks in such a way that when neutron-absorbing control rods made of cadmium were removed, the pile would "go critical," that is, begin a self-sustaining chain reaction. The reaction could be controlled or stopped by reinserting the rods.

On December 2, 1942, the pile was ready to be demonstrated. Under Fermi' close supervision, the last control rods were removed, and the neutron counter indicated that criticality had been achieved. To let the National Defense Research Committee know that the test had been successful, physicist Arthur Compton placed a phone call and uttered the code phrase "The Italian navigator has landed in the New World" ("Italian navigator" was a reference to Fermi). Leslie Groves, head of the Manhattan Project, would describe this event as the single most important scientific event in the development of atomic power.

The nuclear age had begun, but it is remarkable how little regard was paid to safety. This first reactor had no outer shielding, containment vessel, or coolant. Fermi said he was confident his calculations showed that the geometry, or arrangement, of the pile would prevent a runaway chain reaction or an explosion. The "emergency shutdown system" for this reactor consisted of the following: George Weil, who could reinsert the final control rod if necessary; one automatic control rod that was supposed to operate if the radiation intensity reached a preset level; and Norman Hilberry and his axe. Standing on the balcony above the apparatus, if all else failed, Hilberry could use the axe to cut a rope that would release another control rod. If everything else failed, the final resort was a sort of bucket brigade of workers who could dump a neutron-absorbing cadmium salt solution over the pile.

Later, the official historians of the Atomic Energy Commission would note that the "gamble" remained in conducting "a possibly catastrophic experiment in one of the most densely populated areas of the nation!" (U.S Department of Energy 2013).

While such a catastrophe was avoided (one wonders what would have happened to the development of nuclear power in its aftermath), several people would die of radiation in the course of these early experiments. There was also a close call reported by fission pioneer Otto Frisch:

> just as we were getting close to critical size the student who was assisting me pulled out the neutron counter

which he said seemed unreliable. I leaned over, calling out to him to put it back, and from the corner of my eye I saw the neon lamps on the scaler [radiation meter] had stopped flickering and seemed to glow continually. Hastily I removed a few pieces of uranium-235, and the lamps returned to their flickering. Obviously, this assembly had become critical because my body—as I leaned over— reflected neutrons back into it. By measuring the radioactivity in some of the uranium-235 bricks afterwards, we could calculate that the reaction had been growing by a factor of 100 every second! As it happened, I had received only about one standard daily dose [of radiation] in those two seconds, but it would have been a lethal dose if I had hesitated for two seconds longer. (quoted in Wilson 1974, 61)

The willingness to take not fully understood risks to achieve their goal extended to the testing of the atomic bomb itself. It wasn't certain just how hot the center of an atomic bomb blast would get, and at a high enough temperature, nuclear fusion could occur (a feature that would be used a few years later for the hydrogen bomb). Once a fusion reaction got started, might it spread through the nitrogen in the earth's atmosphere and perhaps even the hydrogen in the water of the oceans? (Ironically and perhaps apocryphally, the Nazis—specifically Hitler and the physicist Werner Heisenberg—were having the same conversation. Supposedly even Hitler decided that he didn't want to take a chance on blowing up the whole world.)

Robert Oppenheimer, the theoretical physicist who some would later call "the father of the atomic bomb," was not sure this couldn't happen. He went to Arthur Compton, who was respected for his mathematical prowess. Compton later recalled:

I'll never forget that morning. I drove Oppenheimer from the railroad station down to the beach looking over the peaceful lake. There I listened to his story . . .

Was there really any chance that an atomic bomb would trigger an explosion of the nitrogen in the atmosphere or the hydrogen in the ocean? This would be the ultimate catastrophe. Better to accept the slavery of the Nazis than to run a chance of drawing the final curtain on mankind!

We agreed there could be only one answer. Oppenheimer's team must go ahead with their calculations. (quoted in Rhodes 1986, 418)

The calculations did seem to put the matter to rest. Evidently, any fusion reaction that occurred would be overtaken by cooling and stop by itself. It would take more than one atomic bomb to destroy humankind.

Finally, the "device"—as they called it—was ready to be tested. The Los Alamos scientists waited in the predawn of July 16, 1945. A series of thunderstorms had rolled through, each delaying the nuclear countdown. "And then, without a sound, the sun was shining, or so it looked," Frisch later recalled. Another physicist, Isidor Rabi, recalled of the light that "It blasted, it pounced; it bored its way into you. It was a vision which was seen with more than the eye" (quoted in Gleick, 1992, 154)

Nuclear Perils and Promises

The first time most people heard of nuclear energy was when atomic bombs were dropped on the Japanese cities of Hiroshima and Nagasaki in August 1945. The scale of destruction in the two cities was hard to grasp even after years of wartime bombing campaigns. Within four months of the bombings, at least 150,000 people had died from blast, burns, radiation sickness, or other lingering effects whose significance would be recognized only gradually.

Ever since the industrialization of war began in the nineteenth century, writers had suggested that the latest weapons had made war too terrible to contemplate. This was said of the machine gun and the airplane, yet two devastating world wars had been fought in a generation. What would happen when atomic bombs became part of the arsenals of leading nations? (The Soviets would explode their first atomic bomb in 1949.)

Nuclear weapons were especially frightening because they not only could kill through searing heat and mighty shock waves, they could also kill silently and invisibly—sometimes months or even years later. This was due to radiation—not only the powerful gamma rays emitted at the time of the blast, but the radioactive materials they spewed into the atmosphere in the form of fallout. This aspect of nuclear weapons had been little understood by the scientists of the Manhattan Project. It was only in studying the illnesses suffered by many who had initially survived the Hiroshima and Nagasaki attacks that the full effects of radioactivity on the human body and the environment began to become clear.

Nuclear power was introduced far more abruptly than the earlier forms of energy from coal. Coal involved combustion, a process that had been used since the time of proto-humans at least half a million years ago. Coal itself was used for heating before it became the fuel for steam engines starting in the eighteenth century. Certainly fire was potentially destructive, but its spread and effects were understandable.

Electricity, which mainly involves the electrons in the outer part of atoms, was more mysterious, but its introduction to everyday life was rather gradual, spanning the nineteenth and early twentieth centuries. Neither combustion nor electricity suddenly appeared in the form of an instrument of mass destruction, but this is exactly how nuclear energy entered popular consciousness.

Although the destructive effects of atomic energy aroused great concern and would shape the nature of the ensuing Cold War between the United States and the Soviet Union and their respective camps, another aspect of nuclear energy offered something much more hopeful and positive. It had already been demonstrated that controlled nuclear reactions were possible.

The Atomic Energy Commission

The potential of atomic energy for both destruction and peaceful power generation clearly required an institution that could control the deployment and use of fissionable materials. In 1946 Congress and the Truman administration established the Atomic Energy Commission (AEC). The legislation, called the McMahon/Atomic Energy Act formally transferred control of nuclear materials from the military to a civilian agency. It would be this agency, however, not private companies, who would control research and own any nuclear reactors.

Many of the major research facilities from the Manhattan Project would be incorporated in a new system of national laboratories that would be operated by contractors under AEC supervision. One of the first was Argonne National Laboratory outside of Chicago, which carried on much of the work on reactors begun by Enrico Fermi and his team. Facilities for developing and manufacturing nuclear weapons were also established, such as the Los Alamos Scientific Laboratory (later National Laboratory) in New Mexico, and the Lawrence Livermore National Laboratory in California.

One major task of the new agency was to restrict access to the information and technology involved in creating nuclear weapons. At the time, the United States had a monopoly on atomic weapons, which was felt to be a great military advantage as tensions between the United States and the Soviet Union heated up, particularly in divided Germany. However, in 1949, greatly aided by its effective espionage on the Manhattan Project, the Soviet Union exploded its own atomic bomb.

Reactors at Sea

Interestingly, the first use of nuclear reactors for practical purposes rather than research or fuel production would be as propulsion devices for ships, particularly submarines. This may seem odd, but a look at naval history gives an explanation.

In the age of sailing navies (roughly to the mid-nineteenth century), ships were tethered to the wind. That is, the main limitation on where ships could go and how fast they could get there was the need for favorable winds. When navies switched to coal-burning steam engines and later, oil-burning or diesel engines, ships were pretty much free to set whatever course they wished. However, they were now tethered to the frequent need for refueling. Countries with large ocean-going navies (particularly Great Britain) had to establish a worldwide network of fueling stations.

For submarines, a few other factors also entered in. These underwater marauders were quite effective in World War II. For a time, Germany's U-boats threatened to choke off food and supplies to the island nation of Great Britain. In the Pacific, American submarines proved quite effective against Japanese shipping.

Submarines were most effective when they could stay at sea on patrol for many months. The need for fuel limited their operations. Further, despite their ability to dive, conventional subs are more like whales than like fish, that is, they are "air breathers." They could not run their diesel engines underwater because combustion requires air. They were thus limited to the slower speed and limited duration of electric motors run from batteries. When subs had to surface to recharge their batteries (or to obtain fresh air to breathe), they were vulnerable to enemy destroyers and bombers. (The invention of the snorkel device did allow subs to run underwater while obtaining air, but they had to remain close to the surface.)

Nuclear power offered a supply of fuel that would last for many months. Since nuclear power does not involve combustion, it does not need an air supply. Thus, the use of a nuclear reactor offered the opportunity for a new generation of submarines to undertake lengthy stealth missions anywhere in the world.

On the other hand, the challenges inherent in the project seemed daunting. Any nuclear reactor in a submarine would have to be very reliable, since while it was at sea the ship would be far from any repair facilities. The reactor would have to be able to withstand possible damage from collision or combat. A whole cadre of skilled crew persons would have to be trained to do things that had previously been done only by scientists or engineers at experimental facilities.

These difficulties and the inherent conservatism of many senior military leaders meant that just as advances such as naval aviation required determined leadership, the building of a nuclear submarine fleet found its advocate in Captain (later Admiral) Hyman G. Rickover, the head of the AEC's Naval Reactors Branch. Rickover served with distinction on a variety of ships during World War II. Toward the end of the war, he became more involved in working with electrical systems and gained experience in managing technical projects.

As a leader, Rickover would be controversial. When he appeared on the cover of *Time* magazine in 1954, his wartime efforts were described thusly:

> Sharp-tongued Hyman Rickover spurred his men to exhaustion, ripped through red tape, drove contractors into rages. He went on making enemies, but by the end of the war he had won the rank of captain. He had also won a reputation as a man who gets things done. ("The Man in Tempo 3" 1954)

Rickover found it hard at first to interest his superiors in nuclear propulsion. However, he caught the ear of the chief of naval operations, the famed Fleet Admiral Chester Nimitz, who saw what nuclear power might mean for submarines and other ships.

The prototype reactor designed under Rickover's supervision brought vital experience for the power plants that would follow. This included the design of a reactor that could produce steam and the determination and fabrication of materials that could withstand high temperature and pressure (and, aboard ships, corrosive salt water).

Rickover also instituted a "safety culture" in the design and operation of naval reactors that would be acknowledged even by many later critics of the nuclear power industry. When asked by a colleague why he was so concerned with safety, he replied, "I have a son. I love my son. I want everything that I do to be so safe I would be happy to have my son operating it. That's my fundamental rule" (quoted in Rose 2006, 55).

The first nuclear-powered submarine, *Nautilus*, was commissioned in 1954. Soon the development of huge nuclear submarines that could launch nuclear-tipped intercontinental ballistic missiles (ICBMs) would be a major part of the mutual deterrence strategy of the Cold War. Nuclear power would also prove to be useful for the U.S. Navy's aircraft carriers, starting with the USS *Enterprise*, which was launched in 1960. This huge ship carried eight nuclear reactors. It would be followed by 10 Nimitiz class nuclear carriers. Although there have been conventional mishaps aboard these ships, there have been no significant nuclear accidents.

Other Forms of Nuclear Propulsion

In part because of the success at sea, a number of projects began to work on other applications of nuclear propulsion. One potential application was a nuclear-powered aircraft that could fly thousands of miles without refueling. This project, called NEPA (Nuclear Energy Propulsion, Aircraft) literally never got off the ground. There were two main problems: radiation shielding thick enough to protect the pilot made the plane too heavy to fly, and even if the plane flew, a crash would likely scatter radioactive materials over a wide area. At the same time, the development of long-range missiles (ICBMs) increasingly made long-range bombers obsolete as a delivery platform for nuclear weapons.

Rather similar problems would face designers of nuclear-powered land vehicles. Although companies such as Ford and Studebaker designed nuclear concept cars, no automotive nuclear power plant was ever built.

For a time the space race seemed to offer better possibilities for using nuclear power. The more plausible design is the nuclear thermal rocket, which is rather like a nuclear power plant, except its output is a high-temperature jet of some fluid, generally liquid hydrogen. In theory, a thermal nuclear rocket would be at least twice as efficient (in terms of thrust to weight ratio) as the best chemical rockets. A number of thermal rocket projects were undertaken. The most advanced, under the name Project Rover, was run from 1955 to 1972 and were considered far enough along to build a flying prototype. Some research continues today, but concern about a rocket crash or explosion spreading radioactive materials continues to be an obstacle to this kind of project as well.

Perhaps the most ambitious (if not bizarre) nuclear propulsion project went under the name of Project Orion. The basic idea was to have a rocket eject and detonate a series of small nuclear bombs. The explosions would be shaped to provide maximum thrust; the ships would have a massive "pusher plate" that would absorb the rest of the explosion. The overall performance would be much higher than for chemical or thermal nuclear rockets. It was envisaged that the Orion could be used to deliver people and/or payloads to Mars and other parts of the solar system, and eventually it could even undertake interstellar missions.

Originally Orion was designed to take off directly from the earth's surface, but concern about nuclear fallout was hard to address. The Partial Test Ban Treaty of 1963, which outlawed nuclear explosions in the atmosphere, essentially killed the project.

However, nuclear power systems did find their way into space in a number of applications. The Soviet Union launched a number of satellites with onboard nuclear power reactors. In one case in 1977 the reconnaissance satellite Kosmos 954 went out of control. A system designed to "park" the satellite's nuclear reactor in a safe orbit failed. On January 24, 1978, Kosmos 954 entered the atmosphere over western Canada, scattering radioactive debris. A small portion of the debris was recovered, including one piece that was radiating at a lethal dose rate. Two other Kosmos satellites had similar but less serious accidents. The U.S. space nuclear reactor program, called SNAP, launched only a single satellite, which eventually shed some possibly radioactive debris, although the main reactor core was successfully placed in a long-term parking orbit.

The potential hazards of flying actual reactors near the earth have led most applications to rely on radioisotope thermoelectric generators (RTGs) rather than reactors. These systems use the heat from radioisotope decay (usually plutonium) to power spacecraft. These power plants allow for the operation of probes that are traveling deep into the solar system (to the orbit of Jupiter or beyond, as with the Cassini mission to Saturn, and the Voyager probes that are now leaving the solar system), where the intensity of sunlight is inadequate for solar panels to generate sufficient power. There is also the case of the Curiosity Mars rover (launched in 2012). This car-sized vehicle is too large to be powered by solar power. The onboard RTG is also not affected by night, sandstorms, or the very cold Martian winter.

Because RTGs and smaller radioisotope heater units (RHUs) do contain highly radioactive material, there remains the risk that a failed launch might cause the units to break open and scatter dangerous material. As a result, some nuclear critics have called for an end to NASA's use of these devices.

"Atoms for Peace"

Meanwhile, back on Earth, some Manhattan Project scientists, such as Edward Teller, continued to work on developing more powerful nuclear weapons, including fusion weapons (the hydrogen bomb) that could be 1,000 times more powerful than the bombs that had devastated Hiroshima and Nagasaki. Others, such as Robert Oppenheimer, hoped that the United Nations could take over the stockpile of nuclear weapons and ensure that nuclear technology would be used only for peaceful purposes. (None of these plans went anywhere, in part because of the opposition of the Soviet Union.)

As the United States and Soviet Union became engaged in the Cold War, it became clear nuclear weapons and nuclear submarines were not the only way nuclear energy could be used to gain geopolitical advantage. Nuclear power technology could also be a potent tool for the United States to forge closer ties with its allies and to win over nonaligned countries. More idealistically, nuclear power also appealed to the hope that the technology that threatened the world could be transformed into a means that could bring abundant power to all.

Within the Eisenhower administration, a publicity (some would say propaganda) campaign was developed to promote the peaceful use of nuclear energy. It was intended to reassure America's allies (particularly in Europe) that the United States would use its nuclear stockpile only to deter war, not provoke it. Domestically, the campaign was intended to make the public aware of the potential of nuclear power and encourage investment and public-private partnership in its development.

On December 8, 1953, Eisenhower addressed the UN General Assembly with what became known as the "Atoms for Peace" speech. After talking about the unprecedented dangers that the development of atomic weapons had brought to the world, Eisenhower noted that there could be no victor in a nuclear war and that an effort must be made to turn toward the peaceful use of atomic energy:

> The United States knows that peaceful power from atomic energy is no dream of the future. The capability, already proved, is here today. Who can doubt that, if the

entire body of the world's scientists and engineers had adequate amounts of fissionable material with which to test and develop their ideas, this capability would rapidly be transformed into universal, efficient and economic usage? To hasten the day when fear of the atom will begin to disappear from the minds the people and the governments of the East and West, there are certain steps that can be taken now. (Eisenhower 1953)

Eisenhower then proposed that all governments with stockpiles of fissionable materials begin regular contributions to a pool that would be used to provide material to other countries interested in researching and developing nuclear power. The United Nations would establish an agency to supervise this process.

The Soviets rejected this proposal, but the commitment to the development of nuclear power opened the way to sharing the previously secret information needed to create nuclear reactors. A number of countries could begin to establish their nuclear industries, although this also brought the danger of proliferation—diversion of fissionable materials to the making of nuclear weapons. An international organization for the monitoring and development of nuclear technology would eventually be established in 1957 in the form of the International Atomic Energy Agency (IAEA), an independent agency that reports to the United Nations.

Science Fiction and Cold War Reality

Advocates of nuclear power could point to some powerful arithmetic. A pellet of uranium the size of a thimble can produce energy equivalent to 1,780 pounds of coal; 149 gallons of oil; or 17,000 cubic feet of natural gas. The public was assured by the new promoters of nuclear power that it was only a matter of time before the atomic cornucopia would create a world in which, according to science writer David Dietz:

all forms of transportation will be freed at once from the limits now put upon them by the weight of present fuels. . . .

Instead of filling the gasoline tank of your automobile two or three times a week, you will travel for a year on a pellet of atomic energy the size of a vitamin pill. . . . The day is gone when nations will fight for oil. . . .

The world will go permanently off the gold standard once the era of Atomic Energy is in full swing. . . . With the aid of atomic energy the scientists will be able to build a factory to manufacture gold.

No baseball game will be called off on account of rain in the Era of Atomic Energy. No airplane will bypass an airport because of fog. No city will experience a winter traffic jam because of snow. Summer resorts will be able to guarantee the weather and artificial suns will make it as easy to grow corn and potatoes indoors as on a farm. (quoted in Ford 1982, 30–31)

In a now famous quote, AEC chairman Lewis Strauss suggested in a 1954 speech to a group of science writers that

Our children will enjoy in their homes electrical energy too cheap to meter . . . It is not too much to expect that our children will know of great periodic regional famines in the world only as matters of history, will travel effort-lessly over the seas and under them and through the air with a minimum of danger and at great speeds, and will experience a lifespan far longer than ours, as disease yields and man comes to understand what causes him to age. (Strauss 1954)

Most people even today don't realize that Strauss was refer-ring not to nuclear fission reactors, but the possibility of devel-oping fusion reactors. Still, science writers (and science fiction writers) did become intrigued with the idea that nuclear power might revolutionize every aspect of daily life, even ushering in a utopia.

But as promising as nuclear power might be for the future, there remained many technical challenges before reactors could be reliable enough to serve as public utilities, taking over the job from coal or gas. Further, the economics did not look good from an investment standpoint. It was clear from the start that even if the government did most of the research, nuclear power plants would be capital intensive, that is, they would require a great deal of up-front investment. Recouping that investment might be difficult, given that this was also a time of abundant and relatively cheap fossil fuels.

The early impetus for developing nuclear power therefore came not primarily from economics, but from Cold War concerns. Many American officials and policy makers were worried that other nations—the Soviet Union and even Great Britain, for example—might take the lead as providers of nuclear technology. In a 1953 speech AEC commissioner Thomas E. Murray suggested that the nation was in a "nuclear power race" for which "the stakes were high" (quoted in Walker and Wellock 2010, 2).

The Atomic Energy Act of 1954

The early development of nuclear power would be controlled primarily by the U.S. government, not by private industry. One reason is that data on nuclear fission were at the time closely guarded military secrets. In addition, private companies were reluctant to invest in an expensive, untried technology. Thus, in 1947 the Atomic Energy Commission established the Reactor Safeguards Committee to develop ways to make nuclear power safe and reliable. The AEC also established the Industrial Advisory Group under chairman James W. Parker to investigate peaceful uses of atomic energy. In the early 1950s experimental reactors were used to demonstrate the generation of electricity from nuclear power.

The Atomic Energy Act of 1954 was a key piece of legislation that would establish a new comprehensive framework for government and private cooperation in the development of nuclear power. It had several major effects. The development of a commercial nuclear power industry was declared to be a national goal. To further this objective, the technical data needed to develop commercial reactors would be made available to companies.

The AEC was now given two tasks with regard to nuclear power: promote its commercial development and develop regulations to protect the public and workers from the potential hazards of that technology. As the Nuclear Regulatory Commission (NRC) historian would later note, "Those functions were in many ways inseparable and proved to be incompatible when they were carried out by a single agency" (quoted in Walker 2010, 4).

Such concerns awaited future critics, however. The thrust here came from the prevailing belief in the power of American ingenuity and enterprise to drive almost any worthwhile goal. As Strauss put it, "The Commission's program is directed toward encouraging development of the uses of atomic energy in the framework of the American free enterprise system." Strauss went on to suggest that "competitive economic nuclear power ... would most quickly be achieved by construction and operation of fullscale plants by industry itself." (Walker 2010, 4)

Early Developments and Regulation

In January 1955 the AEC announced a "power reactor demonstration program." This would be a partnership between the government and private companies. The government would direct its new national laboratories to perform research along those lines most relevant to developing commercial reactors. Private contractors would also be solicited to undertake research programs, with the needed fissionable materials provided free of charge from the government stockpile. Previously restricted data would also be made available.

These efforts did drive considerable research toward developing a viable commercial reactor design. However, the fundamental economics—high startup costs, uncertain return—continued to discourage companies from actually undertaking the building of a commercial nuclear plant. There was also the question of safety and risk. There had not yet been a major reactor accident, but as a General Electric (GE) executive noted in 1954, "no matter how careful anyone in the atomic energy business might try to be, it is possible that accidents might occur" (Walker 2010, 6). Although most people in government and industry thought the chance of a catastrophic accident was very low, if one did occur, who would be legally and financially responsible for it?

Normally, the owner of a facility such as a power plant obtains private insurance to cover accidents. The problem was that insurance is designed to provide for events for which a probability can be calculated, not a potential catastrophe whose consequences could not even be accurately estimated. Private insurance companies proved willing to offer up to $60 million insurance for each power plant—an amount that exceeded just about any existing coverage for industry. However, the utilities were concerned that even this much coverage would not be enough in the event of a major accident.

Two major utilities planning nuclear plants—Consolidated Edison and GE—declared that they would not continue without an adequate government plan to deal with costs that exceeded their private insurance coverage. Fearing that the insurance problem would bring the nascent nuclear industry to a halt, the AEC worked with members of Congress to come up with a solution.

The result was the Price-Anderson Nuclear Industries Indemnity Act, which was passed by Congress in 1957. Under this law plant operators would have to first obtain as much insurance as possible from private sources (as of 2011, the amount was $375 million per plant). A fund was established into which reactor operators were required to pay in the event of an accident that exceeded their private insurance limit, up to $111.9 million per reactor (these amounts are adjusted to account for inflation).

This indirect subsidization of the nuclear industry would be challenged by critics. Another aspect of the act that has been criticized is that it does not allow individuals harmed by a nuclear accident to sue for punitive damages. In the case of *Duke Power Co. vs. Carolina Environmental Study Group (1978)*, the environmental group challenged the law on two grounds. They argued that the law did not provide the "just compensation" for victims that is required under the Fifth Amendment to the Constitution and that it violated the Fourteenth Amendment's guarantee of "equal protection of the laws" by treating nuclear accidents differently from other industrial accidents. However, the U.S. Supreme Court denied these claims, ruling that the government's provisions under the law were reasonable for achieving an important economic objective and that the assurance of a fund for paying damages was a "reasonably just substitute" for private lawsuits. Further, the particular economic circumstances of the nuclear industry were a reasonable justification for treating nuclear power differently from other industries.

Meanwhile, regulations would have to be provided to govern these proposed plant operations. The areas to be covered included control over the storage and distribution of fissionable materials for reactor fuel, radiation safety standards, and the required training and qualifications for reactor operators. For the nuclear plants themselves a two-step licensing procedure was established. First, a company that wanted to build a nuclear plant would submit its plans to the AEC. If the proposed plant was deemed to be safe, a construction permit would be issued. After the plant was constructed, the utility would have to submit to another review of its facilities and operating procedures. If those met safety standards, the reactor would receive an operating permit. This regulatory process would be complicated by the fact that there was no general experience with commercial reactors to draw upon and no reactor designs that had proven to be safe and reliable. As a result, each plant proposal would have to be reviewed on its own merits by the AEC and by a panel of outside experts, the Advisory Committee on Reactor Safeguards (ACRS).

With the regulatory and insurance framework now in place, the AEC's efforts to promote nuclear power began to meet with some success. In 1957 the Shippingport Atomic Power Station in Pennsylvania came online. Built using technological lessons learned from building reactors for the navy, it was the first full-scale nuclear plant devoted solely to producing commercial power. (Its output of 60 megawatts was modest by modern standards.) Over the course of its lifetime (it ceased operation in 1982), the reactor was upgraded twice by changing its core. By 1962 six nuclear power plants were in operation in the United States (though four were partially subsidized by the federal government). By the mid-1960s the nuclear industry really hit its stride, with 50 nuclear plants on order, totaling 40,000 megawatts of generating capacity. By now, nuclear plants had practically become commodities that were sold at fixed prices.

Nuclear power seemed to be the wave of the future. Even the Sierra Club, the premier U.S. environmental group, praised nuclear power as a clean energy source and became one of the strongest advocates for building the Diablo Canyon nuclear plant in California. Responding to concerns about fossil fuel pollution, McChesney Martin, chairman of Florida Power and Light (FP&L), said he would never build another coal plant.

Breeding Controversy

A remarkable characteristic of fission reactions is that they can be arranged so that they create more fuel than they consume.

This is possible because some radioactive isotopes in the fuel, although they cannot sustain a fission reaction themselves, can be converted to fissile materials when they are hit by neutrons from the fission reaction. (For example, in all reactors using uranium fuel, some of the uranium-238 is converted into fissile plutonium.) The ratio between the newly created and original fissile material is called the conversion ratio. In a "breeder reactor" this ratio is greater than one, meaning that the reactor actually creates more fuel than it consumes.

Early developers of breeder reactors pointed to their high efficiency and thus greatly reduced need for new sources of uranium. However, breeder reactors are more complicated and expensive to run than conventional reactors, and a boom in uranium mining led to much lower fuel prices than had been anticipated. Breeder reactors have also been criticized on safety grounds. To get the necessary efficiency, breeder reactors have to be run at high temperatures, which requires special coolants such as liquid sodium, which are corrosive and very flammable if they come in contact with air or water. For those breeder reactors that produce plutonium, there is the additional risk that they may be used to produce plutonium for clandestine nuclear weapons.

Most breeder reactors were produced as experimental efforts. The only commercial breeder reactor in the United States was Fermi I, built in Detroit. The risks associated with this aspect of reactor technology aroused some of the first antinuclear efforts, and the state of Michigan fought unsuccessfully to stop construction of Fermi I, which went online in 1963 and reached full power (but not breeding) operation in 1966, only to have a partial meltdown. This near disaster led John G. Fuller to write a book titled *We Almost Lost Detroit* in 1975.

Expansion of Nuclear Power in the 1960s and 1970s

Meanwhile, as the 1960s progressed, nuclear power plants began to be seen as an industrial commodity, much like a coal or gas plant, although more complicated. General Electric offered to build "turnkey" plants—with everything included at one price so that all the utility had to do was (metaphorically) "turn the key" and begin generating power. General Electric's Oyster Creek plant, built for Jersey Central Power for $66 million would be the first of 13 turnkeys built by General Electric, Westinghouse, and two smaller competitors: Babcock and Wilcox, and Combustion Engineering.

Although the companies promised utilities that the cost to operate these plants would be competitive with fossil fuel units, the economics were controversial, and the developers never released their actual cost figures to build the plants. However, the fact that the reactor builders were willing to offer fixed-price turnkey plants created a sense of confidence in utilities. Between 1966 and 1967 utilities placed 49 new orders for reactors—although these had "cost plus" contracts through which developers could recoup unexpected expenses. Unfortunately, such cost overruns proved to be the rule rather than the exception, with plants often coming in at twice the estimated cost. Construction delays also plagued the industry. By the end of the decade, utilities were starting to cancel orders and even plants that were already under construction.

Looking back on the growing pains of the American nuclear industry, social scientist William Beaver has suggested that if the government had largely limited itself to regulating rather than promoting and subsidizing the development of nuclear power, the industry would still have developed, though more slowly, based on the actual demand for power and the inherent economics. "A government nuclear program more in line with the country's innovative traditions, without excessive government stimulation (roughly $7 billion in today's dollars by 1960) would have allowed the technology to become integrated into the American economic system. Decisions would have been made more on the basis of market conditions and less in response to political needs" (Beaver 2011, 409).

The AEC and the Environmental Movement

As nuclear power stations began to be built in more parts of the United States and around the world, concern about their risks and effects on health and the environment also grew. The documented spread of radioactive fallout from the hundreds of atmospheric nuclear weapons tests during the 1950s increased awareness of radiation in the environment. Meanwhile, the growing environmental movement began to look more closely at nuclear power. For a time, environmentalists seemed conflicted about whether nuclear power was a net plus or minus for the environment. On the one hand, nuclear plants did not produce the pollution that fossil fuel (particularly coal) plants did. On the other hand, the growth of nuclear power brought new concerns, such as pollution caused by uranium mining and particularly the "thermal pollution" that resulted from the large amounts of heat transferred from reactor cooling systems to rivers and other bodies of water. This excess heat could kill many species of fish and encourage the growth of plants that could choke the ecosystem.

The AEC came under pressure to do something about these environmental effects. However, since they had nothing to do with radiation, the agency considered them to be outside its jurisdiction. When several members of Congress tried to amend the Atomic Energy Act of 1954 to include these environmental concerns, the AEC opposed them because the new rules would not apply to fossil fuel plants. While an accommodation was reached, public concern about the environmental effects of nuclear power continued, along with a decline in public trust in the effectiveness of the AEC in addressing such concerns.

In 1969 two prominent scientists, John W. Goffman and Arthur R. Tamplin, challenged the AEC's current standards for how much radiation a nuclear plant could release during normal operation. Goffman and Tamplin argued that if everyone in the United States received the permissible radiation exposure, 17,000 additional cases of cancer would occur annually. (They later revised this estimate to 32,000.)

Many radiation experts within and outside the AEC did not agree with these findings. For one thing, they assumed that everyone lived near enough to a nuclear plant to receive the specified dose. The methodology used by Goffman and Tamplin also assumed that there was no "threshold" at which a dose became significant. Rather, they claimed that there was a linear relationship such that even small doses brought increased risk.

Criticism of the AEC continued as the agency was seen as being too narrowly focused and unwilling to cooperate with the new environmental legislation such as the National Environmental Policy Act (NEPA) signed by President Richard Nixon in 1970. When environmentalists brought a lawsuit against the AEC claiming that its oversight of the construction of the Calvert Cliffs nuclear power plant on the Chesapeake Bay in Maryland, a U.S. district court of appeals agreed.

The AEC also became embroiled in a controversy over another problem—what to do with the radioactive waste being produced by the growing fleet of nuclear reactors. In 1970 the AEC announced plans to develop a permanent repository for nuclear waste. They believed that an abandoned salt mine near Lyons, Kansas, would be suitable. However, the actual geology and hydrology of the area had not been studied, and the state geologist of Kansas and other outside scientists criticized the plans. When further studies were made, the site was shown to be unsuitable—and the AEC's stock in public opinion fell still further.

Amidst all these specific issues, the overall philosophical problem with having an agency both promote and regulate an activity kept surfacing.

The Nuclear Regulatory Commission

Despite growing concerns about the industry, the Arab oil embargo and the ensuing energy crisis of 1973–1974 seemed to offer a new reason to embrace and expand nuclear power as a way to reduce dependence on foreign oil. However, even supporters of nuclear power recognized by then that the AEC would first need to be reorganized.

Congress then passed the Energy Reorganization Act of 1974. The AEC would be replaced by two agencies. The U.S. Energy Research and Development Administration would work to promote nuclear and other forms of energy technology. Regulation of the nuclear industry would be handled by the Nuclear Regulatory Commission (NRC). As the NRC historian notes, "The new agency inherited a mixed legacy from its predecessor, marked both by 20 years of conscientious regulation and by unresolved safety questions, substantial antinuclear activism, and growing public doubts about nuclear power" (quoted in Walker 2010, 49).

By separating promotional and regulatory functions, it was hoped that the new NRC would not be overly influenced by industry agendas. However, this commitment would soon be challenged by events.

From Brown's Ferry to Three Mile Island

The seemingly bright future of nuclear power would soon be challenged by events that would raise serious concerns about the safety and environmental effects of this new technology. Some accidents in experimental power plants had occurred in the 1950s and 1960s, but the 1970s would bring the biggest challenge to the U.S. nuclear industry's promise of safe operation. The first major incident came in March 1975, when an electrician at the Brown's Ferry nuclear power plant in Alabama was using a lighted candle to check for air leaks in the cable room, where there were big bundles of electrical cables supplying power to the plant's many systems. He accidentally started a fire that rapidly burned through some of the cables and cut off power to the controls for the safety systems for both of the plant's reactors. The reactors automatically shut down, but the reactor cores continued to produce heat from residual radioactive decay. The water covering the cores boiled away, and in one reactor the core was within four feet of being uncovered, which could have led to the fuel being subjected to a disastrous meltdown.

Fortunately, technicians were able to jury-rig a pump and restore a flow of coolant before that happened, and no significant radioactivity was released to the environment. Plant officials said that this showed how the plant's multiple safeguards could prevent disaster even when a major safety system was disabled. But a senior project manager at the AEC and NRC, Robert Pollard, resigned after the fire to protest the potential dangers not being taken seriously. By the mid-1970s some engineers had also resigned, deciding that they could no longer support nuclear power. But worse was to come.

In the early morning hours of March 28, 1979, a valve in the condenser system in one of the steam generators at the nuclear power station at Three Mile Island, Pennsylvania, failed. This stopped the flow of water through the plant's secondary cooling system. A backup pump automatically started up to keep the flow of water going. Unfortunately, the valves connecting this backup system to the pipes going to the core had been accidentally left closed, blocking the flow of water. To make things worse, someone had left a paper tag on the reactor's control panel that covered up the light that would have told operators about the problem. Without cooling water, the temperature and pressure in the core began to rise rapidly. As designed, in only a few seconds the reactor's pressure relief valve opened to release the excess pressure, and the reactor executed an emergency shutdown or "scram" to stop the chain reaction. So far, despite an equipment malfunction and a human error, the reactor's multiple safety systems had prevented a serious problem.

Unfortunately, as the pressure in the core fell, the relief valve for the main cooling system failed to close, despite a light in the control room indicating that it had done so. Heat began to build up again as the coolant continued to escape as steam from the reactor vessel into the surrounding containment structure. This triggered the emergency cooling system, but the operators, having no indication that the relief valve was open, assumed that there was now too much water in the reactor. They turned the flow in the emergency cooling system down to a trickle. Then, confused but not panicking, they worked for about two hours to try to reduce the water level, not realizing that the water was coming from the overheated core, and they were experiencing a "loss of cooling accident" (LOCA). When the operators tried to reduce the water buildup by turning off the pumps that circulated water in the reactor core, the remaining water boiled away and exposed the core. The steam reacted with the exposed metal cladding of the fuel rods, producing highly inflammable hydrogen gas. Meanwhile, radioactive fission products were escaping into the containment building, setting off radiation alarms. Finally, the operators realized the relief valve was stuck, and they tried to restore the flow of coolant. But they didn't realize the core had been uncovered.

Confusing bits of information began to leak out into the surrounding community. Local officials began to realize something was wrong at Three Mile Island, but they couldn't get understandable answers from nuclear officials. As water from the reactor overflowed holding tanks in the containment structure and operators released steam to relieve pressure, radiation began to escape into the environment. Suddenly, at 2:00 that afternoon, instruments detected a sudden, brief rise in pressure. Unknown to the operators, the accumulating hydrogen in the reactor vessel had created a small explosion.

Later in the day the operators believed they had regained control of the reactor. By then, however, news of the radiation release had panicked people in the surrounding area. Public alarm only increased when nuclear officials began to worry that a further buildup of hydrogen in the reactor might lead to a major explosion that would spew highly radioactive material for miles around. Plans for evacuating the nearly 1 million people living within 20 miles of the plant were considered, but the NRC then determined that there was no danger of an explosion.

About 10 years later, NRC investigators digging down into the ruined reactor discovered that the top of the core had sunk five feet into the reactor vessel. There had indeed been a melt-down, and it was worse than they had anticipated. But the reactor vessel, though cracked, had held despite the heat of 20 tons of melting uranium fuel at a temperature of up to 5,000 °F.

What is the lesson of Three Mile Island? For nuclear critics and much of the public, it seemed clear enough: Nuclear power was dangerous and prone to getting out of control, operators were fallible, and officials were unwilling to give straight answers. Supporters of nuclear power, however, noted that the containment had held. Despite the meltdown, only a relatively small amount of radiation had escaped into the environment. On the other hand, as would happen repeatedly in the history of nuclear power, an accident had revealed an unexpected route to failure. Instead of the rupture of a large pipe, which was the expected cause of a loss of coolant accident, this near disaster had been caused by a combination of relatively minor mechanical malfunctions, human error, and inadequate safety monitoring. As a result, the NRC began to place considerably more emphasis on "human factors" when evaluating plant procedures and performance.

The massive cleanup effort at Three Mile Island cost utilities, their customers, and taxpayers almost $1 billion—and still left questions about the safety of the site because some radioactive material was left behind. Also left behind is a roomful of studies that have failed to answer the question of why Three Mile Island did not melt down completely or blow up from a hydrogen explosion. Probably the most significant fallout from the near-disaster, however, was the effect on public confidence in the U.S. nuclear industry.

A Growing Antinuclear Movement

Broadly speaking, the antinuclear movement embraces opposition to both nuclear weapons and nuclear power. Although our concern here is with the latter, there is considerable overlap between the two areas. The uranium industry is common to both. The same material can be used for reactor fuel, or if enriched, for weapons. The expansion of nuclear power can increase the risk of diversion of nuclear materials (proliferation). Both areas involve government agencies and big corporations, secret or highly technical information, and little connection with ordinary institutions of daily life.

The 1970s saw a series of protests against proposed nuclear plants in Germany, France, and the Philippines. Meanwhile, protests against nuclear weapons testing peaked, with large demonstrations in Australia and activists in small boats trying to stop French tests in the Pacific. By the 1980s what many perceived as a new nuclear arms race between the Americans and Soviets triggered a larger movement, while protests against power plants continued, with 100,000 marching against the construction of the Brokdorf nuclear power plant near Hamburg, Germany.

Exploring the underlying dynamics of these local and international movements, political scientist Herbert Kitschelt noted that the anti-nuclear movements in France, Sweden, the United States and West Germany

> share similar operational objectives, namely to prevent the completion of nuclear power plants under construction, to prevent work from beginning on planned projects and, ultimately, to shut down existing nuclear facilities. . . . In all of the cases, nuclear power conflicts grew from localized, segmented conflicts about specific power plants into national movements and controversies in the same time period, soon after the first energy crisis of 1973–1974 . . .

Kitschelt further noted that the anti-nuclear movements drew primarily from

> "professionals and (public) service sector employees, farmers and property owners in the vicinity of proposed nuclear sites, and young radicals . . ."

Their opponents were

"a pro-nuclear coalition comprised of nuclear scientists,
engineering firms, utilities and promotional or regulatory
state agencies." (Kitschelt 1986, 57–58)

According to another writer who was both a scholar and a par-
ticipant in the antinuclear movement, the nuclear debate had
thus "reached an intensity unprecedented in the history of technology controversies" in the United States and several
European democracies (Falk 1982, 323–340).

Several organizations emerged or became prominent through
this effort, including the Federation of Atomic Scientists (with
its famous "doomsday clock" logo), Physicians for Social
Responsibility, and the militant environmental organization
Greenpeace. Something then happened that would seem to
confirm many peoples' worst fears about nuclear energy.

The Chernobyl Disaster

By the 1980s public opposition to the building of new nuclear
power plants combined with the lower prices for alternative
fuels such as oil and coal had already slowed the expansion of
nuclear power. However, other nations—notably France and
Japan, which have to import their oil—had embraced the atom
as an energy alternative. The Soviet Union, too, had built an
extensive network of nuclear power plants and was also selling
them to countries in Eastern Europe. But events would soon
reveal fatal flaws in the Soviet nuclear industry.

History's worst nuclear accident (rivalled only by the 2011
Fukushima Daiichi disaster) began on the night of April 25,
1986, at the Soviet nuclear power plant in the village of
Chernobyl. The operators were conducting an experiment with
the steam turbine to see whether in the event of a power failure,
the blades could still "coast" on inertia long enough to drive an

emergency cooling pump. The experiments were running behind schedule, so the operators were rushed. They were too quick in throttling power output from the normal service level down to the 25 percent level needed for the experiment. Shutting down a reactor too fast creates products that absorb neutrons and "poison" the core so that it can't be restarted for many hours. As the reactor became poisoned, operators tried to compensate by pulling the control rods nearly completely out to speed up the chain reaction. Even so, they achieved only 6 percent power. Operations under these conditions are very unstable and indeed were forbidden by the operating rules, but the plant's supervisors and operators did not seem to take regulations seriously.

Despite the danger, the supervisors ordered the experiment to continue. At 1:19 a.m., they turned on the water pumps that were to be tested. Because water absorbs neutrons, it made the poison situation worse, and the operators pulled out the rods even further to keep the reaction going. The reactor was now in a very dangerous condition—if there was a loss of water, the neutrons from the now uninhibited fission could send the reaction out of control, doubling the power every second or so.

At 1:21 a.m., the added water flow was stopped, but operators failed to immediately reinsert the control rods. The computer issued a warning that the reactor was now unsafe and should be shut down, but for some reason the operators ignored the message. Meanwhile, the pumps, being driven only by inertia, slowed down and reduced the water flow still further. Now things really got out of control. Less water meant more heat, more heat meant more steam, more steam meant less water—and fewer neutrons being absorbed, leaving more to accelerate the nuclear chain reaction.

Only about three minutes later, operators began to frantically reinsert the control rods, but that system operated more slowly than those in American reactors, and things were getting worse every fraction of a second. Damage to the core from heat began to block the rods so that they couldn't be inserted all the way.

Soon the heat level was 100 times the reactor's design level. At 1:24 a.m., two loud explosions occurred, and a glowing mass flew out of the top of the reactor building. The reactor had blown up—not like a nuclear bomb but like a giant overheated boiler—a boiler filled with deadly radioactive materials. The explosions were followed by fires. Firefighters heroically put out the fires, many receiving lethal radiation exposures in the process. Helicopters dropped neutron-absorbing boron, clay, lead, sand, and dolomite over the top of the building to dampen the radiation. And some of the pilots, too, would die from radiation sickness.

Despite the beginnings of the era of glasnost, or "openness," Soviet authorities reacted in the old Soviet way: secrecy. The first warning of the Chernobyl disaster came when radiation monitors in Sweden, Finland, and other neighboring countries signaled an abnormal level of radiation in the air. Eventually, the Soviet authorities revealed what had happened, and international teams offered assistance in treating the victims of radiation. People in the surrounding area were evacuated, but they have a somewhat higher risk of cancer now and in the future.

The extent of the disaster shocked people around the world and helped solidify opposition to nuclear power, which combined with the anxiety about possible nuclear war that also seemed to be peaking at the time. Nuclear experts in the United States, however, pointed out that the Chernobyl reactor was different from U.S. power reactors in several key respects. Instead of using water as both moderator and coolant, the Soviet reactors used graphite as a moderator. With water, a loss of flow means a loss of moderator that automatically slows down the nuclear reaction by reducing the number of available neutrons. With a graphite moderator a loss of coolant water doesn't reduce the flow of neutrons, and indeed, the graphite is flammable and likely to add to problems by catching on fire. Further, the Soviet reactors had only an ordinary building enclosing them, not a thick reinforced-concrete containment structure as in U.S. power reactors. Finally, U.S. experts noted that the lax behavior of workers and the ignoring of regulations typical at Chernobyl were not typical at U.S. nuclear plants, where unusual experiments could be carried on only after lengthy planning and under close supervision.

In the 1990s the legacy of Chernobyl survived the fall of the Soviet Union. Fifteen other reactors in Russia and the former Soviet countries use the same Reaktor Bolshoy Moshchnosti Kanalnyy, (RBMK) graphite-moderated design as the Chernobyl plant. Even today Russia finds it very difficult to phase out these plants, and though a number of European nations and the United States have offered technical and financial assistance, there has been no comprehensive plan to deal with nuclear safety. One proposal is to replace the nuclear plants with natural gas plants, but this would cost at least $18 billion plus the cost of the gas itself.

Yucca Mountain and Nuclear Waste

In 1982 Congress passed the Nuclear Waste Policy Act, which gave the U.S. Department of Energy the task of siting, building, and operating a national repository for what had by then become about 70,000 metric tons of spent nuclear fuel and high-level (i.e., highly radioactive) waste that was being stored at 121 sites around the nation, mostly near the power plants that had generated the waste.

The general approach was to find a geographic area that was expected to be geologically stable for at least thousands of years and whose geology would essentially prevent contact between the waste and water or other parts of the surrounding environment. After studying 10 sites, the search was narrowed down to one—Yucca Mountain in Nevada (which also happened to be a former nuclear test site and thus was remote from any settled areas).

The original intention was that starting in 1998 the government would begin accepting waste for storage at what was to be a huge underground facility consisting of a five-mile-long (eight kilometers) U-shaped tunnel opening into hollowed-out "alcoves." All together, the government spent about $9 billion on geological research alone, with construction and operation of the facility expected to cost about $90 billion over an expected 100-year lifetime.

As questions resulted in further studies and delays plagued construction, a vigorous opposition also developed to the Yucca Mountain facility. Polls found that about two thirds of Nevadans opposed the facility. In March 2006 the U.S. Senate Committee on Environment and Public Works reviewed the status of the project and issued a white paper titled "Yucca Mountain: The Most Studied Real Estate on the Planet." Its conclusion included:

- Extensive studies consistently show Yucca Mountain to be a sound site for nuclear waste disposal.
- The cost of not moving forward is extremely high.
- Nuclear waste disposal capability is an environmental imperative.
- Nuclear waste disposal supports national security.
- Demand for new nuclear plants also demands disposal capability.
  ("Yucca Mountain" 2006, 3)

However, Congress was losing enthusiasm, and Yucca Mountain was losing funding. (Democratic senator Harry Reid of Nevada, a longtime opponent of the project, had become majority leader.) By 2009 the partially completed facility was essentially shut down. Attention has now shifted to the possibility of establishing regional waste disposal sites that while not as permanent or secure as Yucca Mountain, might provide an alternative to the continued storage of waste in casks near reactor sites. However by 2013 opposition by Native Americans and anti-nuclear activists had halted development of a large high level waste facility on the Goshute Indian reservation in Nevada.

Fukushima: The Nuclear Perfect Storm

Despite continuing concerns and public ambivalence, by the beginning of the twenty-first century, nuclear power seemed poised for at least a modest expansion and possibly what some advocates called a nuclear renaissance. The growing concern about climate change and the need to reduce carbon and other greenhouse emissions associated with fossil fuels seemed to offer a strong new argument for nuclear power.

Japan, lacking significant fossil fuel resources, had invested heavily in nuclear power. Though it is the only country to have suffered a nuclear attack, Japan has also had an interesting cultural relationship to the atom. In the "atoms for peace" era of the 1950s, a revitalized Westernized Japan was seen as a bulwark against the Soviets and communist China. Although many Japanese wanted nothing to do with the power that had destroyed Hiroshima and Nagasaki, Japanese public opinion, aided by sympathetic officials, gradually swung toward nuclear power. (Popular comic strip heroes of the time included Atom, and his sister Uran [short for uranium]. At the same time, though, movie monsters spawned by radioactivity [notably the giant reptile Godzilla] were shown terrorizing helpless cities.)

During the 1970s Japan solidified its commitment to nuclear power in the face of the Arab embargo and soaring oil prices. Eventually 30 percent of Japan's energy would come from nuclear power, compared to 19 percent in the United States.

Like residents of California, the people of Japan are very familiar with earthquakes and the devastation they can cause in only a few moments. On March 11, 2011, a huge earthquake of magnitude 9.0 took place off the coast of eastern Japan. There was extensive damage to thousands of buildings as well as infrastructure such as bridges and roads. However, worse was to come.

The Fukushima Daiichi nuclear power complex consisted of six separate boiling water reactors and was run by TEPCO, the Tokyo Electric Power Company. At the time of the quake, three reactors were off line for refueling or maintenance. As soon as the quake hit, the three active reactors automatically shut down as they were designed to do. With no power coming from the main grid to the nuclear plant, auxiliary power sources automatically came on to maintain the cooling and control systems.

Unfortunately, the earthquake had also generated a huge wave, or tsunami, that crested as high as 130 feet (43 meters). The auxiliary power generators were housed in low-level rooms, so when the water arrived, it flooded and disabled them. Now there was no power for the cooling pumps, and the temperature within the reactors began to rise. (Although shut down, the reactors continued to generate heat from the radioactivity of their fuel.)

As at Chernobyl, workers struggled to prevent further disaster, even at great personal risk. They scavenged batteries from vehicles in the parking lot and tried to hook them up to bring light and power to dark parts of the plant. They rewired electrical panels, jury-rigged pumps, and manually opened valves to try to get water to the overheating reactor cores. However, their efforts could not keep up with the runaway cores.

The only other way to cool the reactors was to flood them with seawater. However, company officials were very reluctant to do this because the salt water would have ruined the reactors and required very costly replacements. Finally, the Japanese government insisted that the reactors be flooded with salt water. By then, however, it was too late. As they overheated and melted, the cores turned remaining water into steam and generated hydrogen gas. When the hydrogen leaked into surrounding areas, there were a series of explosions, damaging the roof or walls of reactor units 1, 2, 3, and 4. The result was a large release of radiation into the atmosphere and surrounding area, including the ocean (via contaminated water).

The explosions and reports of radiation released into the environment caused considerable alarm, which was exacerbated by lack of reliable information. Destruction of surrounding roads made it difficult to bring emergency help to the area.

Eventually power and cooling systems could be partially restored. However, by then large amounts of radioactive sea water and other contamination were present. (The total radiation released was estimated by the Japanese government as being about one sixth that of Chernobyl.) As a precaution, food grown in the area was barred from human consumption. Although no plant workers were directly killed by radiation, it was estimated that more than 300 people had received "significant" radiation doses that in a few cases exceeded the legal limit.

How many people might die or become seriously ill in the future as the result of exposure to radiation from Fukushima? According to Mark Jacobson, co-author of a relevant Stanford University research report, "There's not going to be zero deaths ... It's not going to be tens of thousands either, but it's not a trivial thing." The study estimates the number of cancer cases in a range from conservative (several hundred) to as many as 1,300. Jacobson explains, "That range uncertainty is primarily a function of three things: The dose of radiation received, where the population was concentrated, and figuring out exactly what the population was being exposed to" (quoted in Koebler 2012).

Meanwhile, Japan, a nation that had generated a third of its power from the atom, has shut down all 54 of its nuclear power reactors by late 2013.

An Uncertain Future

Following the Fukushima disaster, tens of thousands of people demonstrated against nuclear power in Tokyo and other Japanese cities, culminating in a Tokyo demonstration that brought 170,000 protesters. At the same time Japan faced a dilemma. As journalist Olga Belogolova recounts, 87-year-old Hiroko Sata recalls what it was like before Japan had nuclear power and families were limited to just a few light bulbs.

> Sata regards the crowd [of protesters] Then she points to the brightly lit high-rise next to her. "If we don't have nuclear plants, how is it going to work without electricity?" Belogolova goes on to observe that "Japan lacks alternate sources of energy that are plentiful and cheap. After 60 years of dependence, the country is economically, historically, and culturally handcuffed to the atom. It has no ready remedy, and even the long-term fixes could break the Japanese economy" (Belogolova 2013).

Despite some electrical shortages in 2012-2013, the Japanese public and government now seem to be resolved to enter a future without nuclear power, with an increasing emphasis on renewable energy.

The political fallout for the worldwide nuclear industry was also significant. Two weeks after the Fukushima disaster, Germany had massive public demonstrations demanding that the nation's nuclear power plants be shut down. In May, German chancellor Angela Merkel announced that Germany would phase out nuclear power. Eight older German reactors had already been shut down. Those remaining are to be decommissioned by 2022. And although Germany had obtained nearly a quarter of its electricity from nuclear power, predicted blackouts did not occur in 2012–2013.

In the early 2010s, nuclear power seemed to have had something of a revival in the American consciousness. A Gallup poll indicated that nuclear power was viewed favorably by 62 percent of respondents—an all-time high. The Obama administration, which had seemed lukewarm toward nuclear power, agreed to include a tripling of loan guarantees for new reactor construction in a comprehensive energy bill in exchange for a cap on carbon emissions.

Just as the nuclear industry seemed to have begun to recover from Three Mile Island, Chernobyl, and unfavorable energy economics, its future would again become a topic of great controversy. Did Fukushima's combination of massive earthquake and tsunami represent a "perfect storm"—a disastrous combination of circumstances that might happen perhaps once in a century? Would some simple measures such as putting emergency power systems out of reach of flooding prevent such disasters in the future? Or would planning always fall short of unexpected and disastrous events?

Naturally nuclear advocates and opponents would draw opposite conclusions. Former nuclear engineer and pro-nuclear activist Rod Adams suggests:

> if Fukushima Daiichi represents the worst nuclear accident in history, then we should be pursuing a rapidly expanding [nuclear] construction program. Compared to the risks imposed by all other energy alternatives, including the risk that would be imposed by not having sufficient quantities of reliable, affordable energy, Fukushima was not so bad. It was, in fact, a dramatic example that the risks imposed by accepting the benefits of affordable, emission-free nuclear energy are well within the "acceptable" range, even in the worst realistic conditions. (Adams 2011)

On the other hand, Jim Riccio, a nuclear policy analyst for Greenpeace, argues that Fukushima represents a "low-probability high-consequence" event. If the consequences are severe enough, the risk will be too high even if the event is quite unlikely to occur. Riccio argues that the Nuclear Regulatory Commission, having been "captured" by the very industry it is supposed to regulate, has been consistently downplaying the risks of meltdowns like those at Fukushima. He goes on to say:

Despite recognized flaws in their risk assessments, government regulators have allowed the nuclear industry to whittle away at regulations intended to protect the public in order to reduce the cost of producing electricity with nuclear reactors. As a result, safety has been compromised. The nuclear bureaucrats have lost sight of their safety mission and instead have weakened nuclear plant regulations to allow reactors to run longer and harder than ever before. (Riccio 2011)

It remains to be seen what lessons the U.S. nuclear industry and regulators will draw from what happened at Fukushima. Tony Pietangelo of the Nuclear Energy Institute insists:

The U.S. response has been comprehensive, well thought through, has been based on science and fact, and it's been heavily vetted with our stakeholders as well as in coordination with the U.S. Nuclear Regulatory Commission. Really on the day after Fukushima occurred, the industry began to take action in terms of walkdowns of our emergency preparedness, particularly the measures put in place after 9/11 to deal with large fires and explosions to ensure their readiness and that the equipment was staged. Any deficiencies found in those walkdowns were put in site corrective actions programs and were all addressed by the end of last year. (quoted in Wheeler 2012)

On a more personal note, NRC chairperson Allison M. Macfarlane recounted her own visit to Fukushima in December 2011:

I was at loss for words. It was awesome in a way, but in the negative sense, to see these villages. Everything is there. The gas stations, little shops but nobody there, and weeds taking over parking lots. A nice garden with dry dead stuff. Weeds growing over the train tracks.

At the nuclear site, she surveyed tsunami debris and "the rusting carcasses of overturned trucks." Macfarlane said the lesson for the NRC might be to raise the level of the urgency it gives to considering rare events, including tornado hazards. "You might fold climate change into this," she said (quoted in Mufson 2013).

One thing is certain: nuclear power will continue to play an important role both in energy generation and in the debate over our energy future.

References

Adams, Rod. 2011. "Fukushima Happened, Now What?" *Atomic Insights* (December). http://atomicinsights.com/ 2011/12/fukushima-happened-now-what.html

Beaver, William. 2011 "The Failed Promise of Nuclear Power." *Independent Review* (Winter).

Belogolova, Olga. 2013. "Why Japan Can't Quit Nuclear Power." *National Journal* (February 14). Available online at http://www.nationaljournal.com/magazine/why-japan-can -t-quit-nuclear-power-20130214

Da Andrade, E. N. 1964. *Rutherford and the Nature of the Atom.* New York: Doubleday.

Einstein, Albert. 1939. Letter to President Franklin D. Roosevelt. www.hypertextbook.com/eworld/einstein.shtml

Eisenhower, Dwight D. 1953. "Atoms for Peace." http://www .eisenhower.archives.gov/research/online_documents/atoms _for_peace/Binder13.pdf

Falk, Jim 1982. *Global Fission: The Battle over Nuclear Power.* New York: Oxford University Press.

Ford, Daniel. 1982. *The Cult of the Atom: The Secret Papers of the Atomic Energy Commission.* New York: Simon and Schuster.

Gleick, James. 1992. *Genius: The Life and Science of Richard Feynman.* New York: Vintage.

Grady, Sean M. 1992. *The Importance of Marie Curie.* San Diego: Lucent.

Hewlett, Richard G., and Oscar E. Anderson. 1962. *The New World: A History of the United States Atomic Energy Commission. Vol. 1, 1939–1946.* University Park: Pennsylvania State University Press.

Inam ur Rehman. 1993. "Historical Evolution of Nuclear Technology." *Economic Review* (June).

Kitschelt, Herbert B. 1986. "Political Opportunity Structures and Political Protest: Anti-Nuclear Movements in Four Democracies." *British Journal of Political Science*, vol. 16. http://www.marcuse.org/harold/hmimages/seabrook/ 861KitscheltAntiNuclear4Democracies.pdf

Koebler, Jason. 2012. "Report: Fukushima Radiation Could Kill More than 1,000." *U.S. News and World Report* (July 17). http://www.usnews.com/news/articles/2012/07/ 17/report-fukushima-radiation-could-kill-more-than-1000

"The Man in Tempo 3." 1954. *Time* (January 11).

Mufson, Steven, 2013. "In U.S., Nuclear Energy Loses Momentum amid Economic Head Winds, Safety Issues." *Washington Post* (March 11).

Rhodes, Richard. 1986. *The Making of the Atomic Bomb.* New York: Simon and Schuster.

Riccio Jim. 2011. "Fukushima Disaster Shows That Nuclear Power is Never 'Safe.'" *U.S. News and World Report* (July 1). http://www.usnews.com/opinion/blogs/on-energy/2011/

07/01/fukushima-disaster-shows-nuclear-power-is-never
-safe

Rose, Lisle A. 2006. *Power at Sea. Vol. 3, A Violent Peace,
1946–2006*. Columbia: University of Missouri Press.

Sime, Ruth Lewin. 1997. *Lise Meitner: A Life in Physics*.
Berkeley: University of California Press.

Strauss, Lewis L. 1954. "Remarks Prepared for Delivery at the
Founders' Day Dinner, National Association of Science
Writers" (September 16).

U.S. Department of Energy Office of History & Heritage
Resources. 2013. "The Manhattan Project: An Interactive
History." http://web.archive.org/web/20101014114659/
http://www.cfo.doe.gov/me70/manhattan/cp-1_critical.htm

Walker, J. Samuel, and Thomas R. Wellock. 2010. *A Short
History of Nuclear Regulation, 1946–2009*. Washington,
D.C.: U.S. Nuclear Regulatory Commission.

Wells, H. G. 1914. *The World Set Free: A Story of Mankind*.
London: Macmillan.

Wheeler, Brian. 2012. "Special Report: Nuclear Power
Executive Roundtable; Analyzing the Future of Nuclear
Power Generation in the U.S." *Power Engineering*
(October 12). Available online: http://www.power-eng.com/
articles/2012/10/special-report-nuclear-power-executive
-roundtable.html

Wilson, Jane (ed.). 1974. *All In Our Time: The Reminiscences of
Twelve Nuclear Pioneers*. Chicago: Bulletin of the Atomic
Scientists.

"Yucca Mountain: The Most Studied Real Estate on the
Planet." 2006. U.S. Senate Committee on the Environment
and Public Works, http://epw.senate.gov/repwhitepapers/
YuccaMountainEPWReport.pdf

In its more than half a century of existence, nuclear power has raised many important questions. The purpose of this chapter is not to draw final conclusions or make recommendations, but to provide the reader with an overview of the many factors involved in any decision to expand, maintain, or phase out this type of energy production.

We will begin with a discussion of the basic costs involved in building and operating a nuclear power plant, and how the cost of electricity generated from nuclear power compares with fossil fuels and renewable forms of energy. Such a comparison will turn out to be far from simple because of the influence of global markets and the possible effects of government tax and regulatory policies.

Next, we will look at considerations that are unique to nuclear power, such as radiation exposure, other environmental effects, and the management of radioactive waste. Another important consideration for the routine operation of nuclear plants and waste facilities to be discussed is that of nuclear proliferation, or the acquisition by rogue governments or terrorist groups of material that could be used to make nuclear weapons.

The safe disposition of nuclear waste, stored in sites such as this low level waste trench in Hanford, Washington, remains challenging and controversial. (Courtesy of NRC.gov)

This will be followed by a discussion of the safety of nuclear plants, the types of accidents that have occurred, and efforts to improve safety through new reactor designs.

Finally, we will look at two major arguments for the expansion of nuclear power: the search for energy independence and especially, the need to reduce the use of fossil fuels that contribute to climate change.

Nuclear Power Today

As of 2014 436 nuclear power reactors were in operation in 30 countries, according to the Nuclear Energy Institute. Nuclear power provided about 12.3 percent of the world's electricity.

What is the broad trend in the use of nuclear power worldwide? According to data from the 2013 World Nuclear Status Report, total nuclear electricity generation actually dropped by 4 percent in 2011 (the year of the Fukushima disaster) and then dropped a further 7 percent in 2012.

In 2011 the number of reactors fell from 441 to 433, and total generating capacity fell to 366.5 GW (gigawatts, or billions of watts). While 5.1 GW of new capacity was connected to the grid in China, India, Iran, Pakistan, Russia, and South Korea, this was more than offset by the shutting down of about 11.5 GW of capacity in France, Germany, the United Kingdom, and particularly Japan, as a result of the Fukushima disaster.

To look at a longer time span, in 1993 the share of global electricity production from nuclear power peaked at 17 percent, but it had dropped to 11 percent by 2011. Another important current trend involves construction of new reactors. In 2010 construction began on 16 reactors, 10 of which were in China. However, in 2011 only two new reactors began construction—one each in India and Pakistan. In 2012 three completed reactors started electricity generation, but six were shut down. Overall, new facilities coming online have not been keeping pace with shutdowns. According to a report for the Worldwatch Institute, just to maintain existing generation levels as older reactors go out of service, new reactors generating 18 GW would have to be operating by 2015 and another 175 GW by 2025. Recent events have cast doubt on whether such a growth rate is likely.

Most broadly, there are two considerations that will determine the desirability of nuclear power compared to other forms of energy: cost and risk.

How Much Does Nuclear Power Cost?

To compare nuclear power to other forms of electricity generation, it is necessary to look at several cost components. First, there is the capital cost of building and maintaining the power plant and its associated infrastructure. To this must be added the cost of raising the required capital, which includes a "discount rate" reflecting the need to give the investment a sufficiently attractive return to get people to buy bonds or equities. Typically this is about 3.5 percent, but to the extent a technology such as nuclear is perceived as having higher risks, an additional "risk premium" may need to be factored in. As of the 2010s, costs for a large conventional nuclear plant are estimated at $6 to $8 billion.

Capital costs are generally lowest for fossil fuel (coal and oil) power stations and high for nuclear plants. As journalist Daniel Indiviglio notes:

> One of the big problems with nuclear power is the enormous upfront cost. These reactors are extremely expensive to build. While the returns may be very great, they're also very slow. It can sometimes take decades to recoup initial costs. Since many investors have a short attention span, they don't like to wait that long for their investment to pay off. (Indiviglio 2011)

The capital cost also depends on the type of reactor chosen. A University of Chicago report notes, "To the extent new reactor designs are pursued for their advantages such as safety and flexibility in scale, such innovative reactors might increase capital costs up to 35 percent for these new designs" (University of Chicago 2004).

The renewed concern about safety in the wake of the Fukushima disaster may also translate into higher capital costs, as noted by one journalist:

> The proven dangers of nuclear power amplify the economic risks of expanding reliance on it. Indeed, the stronger regulation and improved safety features for nuclear reactors called for in the wake of the Japanese disaster will almost certainly require costly provisions that may price it out of the market. (Folbre 2011)

Overall, nuclear plants have been becoming more expensive to construct in recent years. One analyst found that "Estimated construction costs (exclusive of financing) per one installed kilowatt of production capacity have jumped from a little over 1,000 dollars in 2002 to well over $7,000 in 2009" (Sokolowski 2010). In updating a 2003 Massachusetts Institute of Technology (MIT) study on the future of nuclear power, the authors found in 2009 that the costs of constructing a nuclear plant had approximately doubled (Deutch et al. 2009).

Fuel Costs and Comparisons

Once a plant has been built, there are, of course, ongoing operating costs, of which the most variable is the cost of fuel. Fuel costs are relatively high for fossil fuel, although natural gas prices have been declining and are expected to continue to do so. This is largely because of hydraulic fracturing (or fracking), a technique that makes many new gas fields economically accessible, including fields in North America.

Nuclear power has relatively low fuel costs (partly reflecting the highly concentrated energy in uranium). However, while the cost of obtaining uranium is relatively low, the radioactivity of nuclear waste and the need to keep it secure imposes high costs, as does the eventual decommissioning of nuclear plants that have reached the end of their design life or economic viability.

In 2007 energy analysts with MIT noted:

> Incorporating all cost elements, we find that the levelized cost of electricity from nuclear power is 8.4¢/kWh, denominated in 2007 dollars. The levelized cost of electricity from coal, exclusive of any carbon charge, is 6.2¢/kWh, denominated in 2007 dollars. The levelized cost of electricity from gas, exclusive of any carbon charge, is 6.5¢/kWh, denominated in 2007 dollars. (Du and Parson, 2007)

("Levelized" cost refers to the price at which electricity must be generated to break even over the life of the plant. It is important to note that the charges for electricity from fossil fuel sources do not take into account any "carbon tax" that might be imposed in the future.)

Looking ahead, the U.S. Energy Information Administration (EIA) estimated the cost of generating electricity from new plants that would enter service in 2017 (see Table 2.1).

This and other cost data and projections suggest that among fossil fuels, coal is most expensive (and roughly comparable to nuclear).

Growth Projections

Looking at trends projected to 2040, the EIA notes:

> Coal remains the largest energy source for electricity generation throughout the projection period, but its share of total generation declines from 42 percent in 2011 to 35 percent in 2040. Market concerns about [greenhouse gas] emissions continue to dampen the expansion of

Table 2.1   Nuclear Power around the World

Plant Type	Total System Levelized Cost in Dollars per Megawatt Hour
Conventional coal	99.6
Advanced coal	112.2
Advanced coal with carbon capture and sequestration (ccs)	140.7
Conventional natural gas	68.6
Advanced natural gas	65.5
Advanced gas with ccs	92.8
Conventional combustion turbine	132.0
Advanced combustion turbine	105.3
Advanced nuclear	112.7
Geothermal	99.6
Biomass	120.2
Wind	96.8
Solar photovoltaic	156.9
Solar thermal	251.0
Hydroelectric	89.9

(Energy Information Administration 2012)

coal-fired capacity ... even under current laws and policies. Low projected fuel prices for new natural gas–fired plants also affect the relative economics of coal-fired capacity, as does the continued rise in construction costs for new coal-fired power plants. (Energy Information Administration 2013)

The EIA notes that prices for natural gas are lower than previously projected, and "New natural gas–fired plants are also much cheaper to build than new renewable or nuclear plants."

Renewables have a more mixed profile. Hydroelectric, the most widely used renewable (accounting for 16 percent of electricity generated worldwide in 2010), is also the least expensive. However, very few new hydroelectric dams have been built in the United States in recent years. This is due both to the limited number of additional suitable sites and concern about the environmental effects of damming rivers.

Among the other renewable sources, geothermal is cost effective in only a few locations. Wind is available in more places, but large wind installations can have deleterious effects on land, scenery, and birds. There have been some promising developments in solar technology, but for now solar is the most expensive option for large-scale power generation.

The EIA's growth projection for nuclear power is rather modest—from 790 billion kilowatt hours in 2011 to 903 billion kilowatt hours in 2040. At that time nuclear would account for 17 percent of total generation, down from 19 percent in 2011. These figures are based on current plans for new plants as well as upgrades in capacity of existing plants and removing older plants from production.

The share of power from renewable sources (hydro, solar, and wind) is projected to grow from 13 percent to 16 percent during the same period. However, the EIA notes that this projection does not take into account the renewal or possible expansion of existing subsidies and tax breaks for renewable energy, or, for that matter, policies such as a carbon tax that could make electricity derived from fossil fuels considerably more expensive.

Subsidies, Incentives, and Regulations

The economics of virtually every form of energy are complicated by subsidies and tax credits of various kinds, as well as current or potential regulations. However, this seems particularly true of the nuclear industry. Charles Ferguson notes that "nuclear power has never succeeded anywhere without enormous government backing" (Ferguson 2011b, 51). He cites France and China as examples. However, Ferguson also notes that oil is also heavily subsidized, both through tax breaks and indirectly because protecting access to foreign oil is a major task of the enormous U.S. military.

One subsidy that has long been a point of contention for nuclear critics is the insurance guarantees under the

Price-Anderson Act, which serve as a sort of "catastrophic coverage" for the industry. However, as of 2013 this coverage had never been used—even the largest claims, amounting to $71 million for the Three Mile Island accident, were paid from the utilities' regular insurance policies. However, this leaves open the possibility that a much large disaster—perhaps on the scale of Fukushima, if not Chernobyl—would result in American taxpayers being on the hook for tens of billions of dollars worth of claims. As a result, critics question whether the benefits of nuclear power justify taking this risk, while proponents argue that such coverage of remote risks was a necessary and reasonable way to allow an important industry to develop.

While the insurance subsidy thus is indirect and somewhat hypothetical, the nuclear power industry (and other forms of energy as well) benefits directly from a variety of tax incentives. For example, the Energy Policy Act of 2005 included the following incentives:

- A tax credit of 2.1 cents per kilowatt hour for the first 6,000 megawatts of new nuclear capacity, for the first eight years of a plant's operation
- Insurance of $2 billion to cover costs from regulatory delays for full power operation of the first six advanced new plants
- Federal loan guarantees for advanced nuclear reactors, up to 80 percent of the project cost
- A 20-year extension of the Price-Anderson Act insurance and liability protection
- More than $2 billion for research in advanced high-temperature reactor development, including the ability to co-generate hydrogen fuel (this became part of the Nuclear Power 2010 Program)

The question of tax policies brings up a significant problem with comparing energy costs. Every form of energy is affected to a considerable extent by the presence of tax breaks on the one hand and potentially expensive regulations or taxes on the other. For example, tax credits or loans can make it more attractive to build solar panels or wind turbines.

However, these efforts pale besides the subsidies that have been provided for producers of fossil fuels, particularly oil. Originally these subsidies (generally in the form of tax breaks) were intended to offset some of the risk involved in exploring or drilling for oil. It can be argued, however, that improvements in technology and the increasingly global nature of the oil market have made the subsidies both less necessary and less effective. Thus, "the economic analyses of the impact of coal and gas subsidies show very little response in domestic production" (Aldy 2013).

On the other hand, taxes on carbon emissions (which have been proposed but not enacted at the federal level in the United States) would make production of energy from fossil fuels more expensive. This would tend to tip the balance toward power generation that does not result in carbon emissions such as renewable (water, wind, and solar) but also nuclear.

What would be the effect of not continuing current subsidies for constructing new nuclear plants? The Chicago report, looking at the levelized cost of electricity (not the same calculation as in Table 2.1), concludes:

> Without federal financial policy assistance, new nuclear plants coming on line in the next decade are projected to have a levelized cost of electricity (LCOE) of $47 to $71 per megawatt-hour (MWh). This study provides a full range of LCOEs for first nuclear plants for alternative construction periods, plant lives, capacity factors, and overnight cost estimates. LCOEs for coal- and gas-fired electricity are estimated to be $33 to $41 per MWh and $35 to $45 per MWh, respectively. (University of Chicago, 2004)

Although these calculations express a considerable range of uncertainty, they seem to suggest that the economic viability of nuclear power is greatly influenced by favorable policies (tax credits, accelerated depreciation of equipment, investment credits, production credits). This is unlike the case noted earlier with subsidies for fossil fuel production, which seem to have little effect on production.

The other side of the equation is that policies that make electricity from fossil fuels more expensive—such as more stringent emissions limits, a carbon tax, or required carbon sequestration—would give both nuclear and renewable power a stronger competitive position. Thus, the Chicago study notes:

> If stringent greenhouse policies are implemented and advances in carbon capture and sequestration prove less effective than hoped, coal-fired electricity's LCOE could rise as high as $91 per MWh and gas-fired electricity's LCOE could rise as high as $68 per MWh. These LCOEs would fully assure the competitiveness of nuclear energy. (University of Chicago 2004)

Continuing Uncertainty

The sensitivity of nuclear power to energy economics and the regulatory environment makes it difficult to make any real predictions about what will happen to the competitive position of nuclear power in the next few decades. About the best one can do is identify the factors that seem to be more relevant and watch them closely.

In December 2010 a conference brought together government officials and leading nuclear industry figures. Chip Pardee, the chief nuclear officer of Exelon, a leading operator of nuclear reactors, said, "We can't make the numbers work" with the existing level of federal aid (quoted in Wald, 2010).

At the same meeting U.S. secretary of energy Steven Chu was described as believing that

> while new reactors were likely to be important industrial assets for 60 or 70 years, the market was focused on the

short term. Low prices for natural gas, a competing fuel, and the collapse of efforts to impose a price on carbon dioxide emissions make the economic climate for new reactors quite unfavorable at the moment. (quoted in Wald, 2010)

Chu noted that while the Obama administration had been trying to expand federal loan guarantees for the industry, even the existing funds were not being fully utilized. He went on to say, "In addition to loan guarantees, you need an environment that's right, that makes it look like a good investment." (quoted in Wald, 2010)

As for the prospects of a carbon tax that would make nuclear a stronger competitor to fossil fuels, presidential energy advisor Carol Browner agreed that "A price on carbon would be hugely beneficial to this industry." However, she went on to note that thus far a "carbon cap" has not made it through Congress and that it "may not be possible in the near term." (quoted in Wald, 2010)

The only positive note for the nuclear industry at the conference was suggested by a former member of the Nuclear Regulatory Commission, James K. Asselstine, now the managing director of high-grade credit research at Barclays Capital:

Demand for electricity had been growing by about 1 to 1.5 percent a year, but has actually fallen in the last two years, so it is now about 8 percent below where it would have been had no recession emerged. . . .

And in the next year, a large number of old coal-fired plants are likely to be retired because of pending rules from the Environmental Protection Agency on the emission of pollutants like sulfur dioxide, soot and mercury. When that coal-fired capacity is gone, . . . nuclear's prospects might look brighter. (quoted in Wald, 2010)

The 2011 Fukushima disaster may also be playing a role in economic perceptions of nuclear power. According to

Timothy Frazier of the Department of Energy's Blue Ribbon Commission on America's Nuclear Future:

> Fukushima did not stop the "nuclear renaissance" in the United States. Fukushima did, however, remind people in the United States of the potential impact of an incident at nuclear power plants. The nuclear renaissance in the United States is being slowed by the historically low price of natural gas, which is being used to generate cheap electricity. (quoted in Wright 2012)

However, Peter Bradford, formerly a member of the NRC and now with the Union of Concerned Scientists, counters this assessment, saying:

> The nuclear renaissance was on the rocks well before Fukushima, as industry leaders like John Rowe had acknowledged publicly. The reasons were economic. Fukushima has made it harder for the nuclear industry to justify the economic subsidies that it must have to off-set the economics. It has raised costs for the operating reactors. It has affected public opinion ... (quoted in Wright 2012)

Radiation and Health

Moving from the cost side of the nuclear equation, we come to the area of risk. Despite the great enthusiasm for nuclear energy expressed by its early advocates, the danger inherent in the radiation that it depended on could not be ignored. Radiation can make people sick and sometimes kill them, either immediately or years later through cancer. It can also cause genetic mutations that lead to crippled offspring. Yet radiation is not really exotic: It is a pervasive part of the natural environment.

Radiation doses are measured in rem (radiation equivalent man) or millirem (a thousandth of a rem). On the average, people are exposed to about 360 millirem per year from natural background radiation. This radiation comes from cosmic rays penetrating the atmosphere from space as well as from small amounts of radioactive material in the rocks, soil, and building materials that surround us.

Several factors can influence the amount of background radiation a person receives. People living in Colorado, for example, receive about 70 millirem more in background radiation than those living in Florida because the higher altitude in Colorado means that there is less atmosphere to shield people from cosmic rays. (This also explains why crews on commercial airliners actually have a higher annual radiation exposure than workers in nuclear power plants.) People living in houses built on soil that produces the radioactive gas radon from decaying traces of uranium can also have greater-than-average exposure, for the gas tends to accumulate in basements.

People receive artificial radiation from a number of sources— mainly medical X-rays and cancer treatments, but also consumer products such as smoke detectors and clocks with illuminated paint on their dials. Compared to natural sources, a nuclear power plant that operates without accident adds very little to the radiation already in the environment. Typically it exposes people in the surrounding 50 miles to only an additional .01 millirem per year.

Radiation can damage living tissue in several ways. High doses of 100 rem or more physically disrupt and destroy cells, leading to radiation sickness with symptoms such as vomiting, intestinal bleeding, and anemia (destruction of red blood cells). Half of all persons exposed to 400 rem of radiation will die. Smaller doses may have no immediate effect but can lead to an increased risk of leukemia or other forms of cancer. Finally, radiation exposure of the reproductive cells can damage the DNA sequences, leading to mutations whose effect on offspring can range from insignificant to lethal.

Given these potential effects, and given the amount of radiation already present in the environment, how much additional human-made radiation is too much? One theory says that small doses below some as-yet-unknown threshold have no significant health effect. However, most scientists now believe that any dose of radiation, even a small fraction added to the background dose, introduces some additional risk. Recent research suggests that this "linear" relationship between dose and effect is more likely to be correct, at least until the dose becomes very small.

The study of "cancer clusters," geographical areas that show an unexpected number of cases of leukemia (cancer of the blood), has further fueled the controversy. In the 1980s cancer clusters were discovered in the vicinity of the Pilgrim nuclear power plant in Massachusetts as well as near several plants in Great Britain. Nuclear critics argue that these clusters arise from the effects of added radiation exposure from the plants, but it has proven difficult to determine whether the clusters are statistically significant and, if so, whether they really correlate to the small amounts of radiation emitted by the plants. (Paradoxically, at Pilgrim, the largest cluster turned out to be in the study area farthest from the plant.)

Further, risk cannot be assessed in a vacuum. Life has hundreds of quantifiable risks, and that leads to difficult trade-offs. Physics professor Richard Wolfson points out that "you might consider moving out if a nuclear power plant were built in your neighborhood. On the other hand, suppose your new location requires that you drive 10 miles further in your round-trip commute to work each day. In 10 years of 250 working days each, the risk associated with the extra driving will give an average life-span reduction of nearly 7 days—more than triple the risk of staying next to the nuclear plant" (Wolfson 1993, 73, 75).

Environmental Effects

Most coverage of nuclear power focuses on the risk of accidents, particularly those that result in significant or even catastrophic release of radioactivity into the environment. However, the routine operation of nuclear plants also has ongoing environmental effects. As part of processing the waste stream of liquid and gas through evaporation or filtering, most but not all radioactive material is removed. The amount of radiation that can be released into the environment is limited by regulation. A typical dose for someone living within 50 miles of a nuclear plant is 0.1 microsievert pert year. This is less than a thousandth of the typical level of natural background radiation. Radiation emissions must be continually monitored, and mathematical models are used to predict the likely dose.

There are also small unplanned leaks that can be considered to be somewhere between routine emissions and accidents. A recent concern has been the detection of tritium (a radioactive isotope of hydrogen) that has become dissolved in water that eventually becomes part of the drinking water supply.

Although "normal" plant emissions seem to be very small, there have been attempts to calculate their contribution to the risk of cancers such as leukemia. A 2007 meta-analysis (essentially a study of studies) by Baker and Hoel suggested higher than normal rates of leukemia among children living near nuclear facilities in a number of countries, including the United States, Canada, the United Kingdom, Germany, and Japan (Baker and Hoel 2007). This study has been criticized for using inconsistent definitions and methodology. A 2008 German study also found elevated leukemia rates. However, many other studies have failed to find to find an association of nuclear plants with higher cancer rates, including a British study (Elliott 2011).

When considering routine emissions, however, nuclear plants compare quite favorably to coal-fired plants. The "fly ash" produced by burning coal has been shown to carry into the environment about 100 times the radioactivity produced from the equivalent amount of energy generated in a nuclear plant. This is because coal contains traces of uranium and thorium. Although tiny, these traces are concentrated in the ash.

It has been estimated that people living near a coal plant may receive up to three times the dose received by those whose neighbor is a nuclear plant.

In absolute terms, however, the radiation risk to health from either type of plant is very low, as noted by Dana Christensen of Oak Ridge National Laboratory: "You're talking about one chance in a billion for nuclear power plants. And it's one in 10 million to one in a hundred million for coal plants" (quoted in Hvistendahl 2007). Indeed by far the biggest health effects of coal plants come not from radiation but from chemical pollution (such as sulfur dioxide, nitrogen oxides, and mercury) and particulates.

What about the environmental and health impact of other forms of energy generation? A 2005 European Union study called ExternE ("Externalities of Energy") found that nuclear energy had a lower economic cost of health and environmental effects than some forms of renewable energy, including biomass and solar photovoltaic cells (due to environmental effects of their manufacturing process). Wind power, however, came in at half the environmental cost of nuclear power (ExternE 2006).

Besides possible effects on human health, nuclear plants have other effects on the surrounding environment. About 60 to 70 percent of the thermal energy generated in a nuclear plant must ultimately be exchanged or dissipated into the environment, either through a body of water or by evaporation in cooling towers. (Nuclear plants are less thermally efficient than coal plants and thus create more waste heat.)

The problem with the waste heat is that it changes the ecological balance of surrounding bodies of water such as rivers or lakes. This can kill existing species of fish or other aquatic life and lead to other undesirable effects such as overgrowth of algae or other organisms that thrive in the warmer water.

The Nuclear Fuel Cycle

Even if a nuclear plant operates perfectly with only minimal release of radiation to the environment, there is still the problem of what to do with radioactive waste. Understanding the problem of nuclear waste begins with a look at how nuclear fuel is made and used, and what happens to it after it is no longer usable in the reactor.

First, uranium must be mined. In the United States, uranium is mined mainly in the western and southwestern states—but increasingly it is imported from places such as Kazakhstan, Canada, and Australia. The uranium is obtained either by the ore being dug out and milled (crushed and ground) or by the uranium being leached, or dissolved within the rock and percolated up and collected. This latter method reduces environmental impact by minimizing the accumulation of tailings (leftover ore that contains some radioactive uranium). The uranium is shipped from the mine in the form of a powder called yellowcake. (A ton of ore yields about four pounds of yellowcake.)

Uranium emits radioactive radon gas as a decay product. While radon can be a concern in areas where it accumulates in homes or buildings as a result of uranium in nearby rock or soil, radon is a particular health concern for uranium miners. In the late 1940s and 1950s, uranium miners in the Four Corners region of the American southwest experienced a significantly increased incidence of lung cancer. In later years strict standards required extensive ventilation and monitoring of radon levels.

Uranium mining has also left a legacy of large contaminated sites, particularly in parts of Utah, Colorado, New Mexico, and Arizona. Hundreds of abandoned mines still await cleanup. Particularly affected are many members of the Navajo Nation of Native Americans. In 2007 the Environmental Protection Agency, the Navajo Nation, the Bureau of Indian Affairs, the Nuclear Regulatory Commission, the Department of Energy, and the Indian Health Service developed a five-year plan to deal with uranium contamination. There are also ongoing efforts at the state level.

To continue with the fuel cycle: the yellowcake goes to a conversion plant where it is chemically changed into uranium hexafluoride gas, which is shipped in cylinders to an enrichment plant. (The three enrichment plants in the United States are in Oak Ridge, Tennessee; Paducah, Kentucky; and Portsmouth, Ohio.) The enrichment plant uses a process called gas diffusion to separate out the lighter, faster-moving particles of the fissionable uranium-235 from the heavier, nonfissionable uranium-238. This raises the concentration of uranium-235 from its natural level of less than 1 percent to 3.5 percent, which is enough to sustain the fission chain reaction in a nuclear power plant.

The enriched uranium is then fabricated into fuel by being compressed into pellets about the size of a fingernail. The pellets are inserted into zirconium tubes (rods) about 12 feet long. The rods are capped, sealed, and stacked to form a reactor core consisting of about 50,000 rods and weighing about 100 tons.

As the fuel undergoes nuclear reaction to generate power, it gradually accumulates fission products—the highly radioactive isotopes that result from the uranium atoms being split. Eventually the presence of these products reduces the efficiency of the nuclear reaction to the point where it can no longer generate adequate power. The material in the core is now considered to be "spent fuel." It contains radioisotopes such as cesium-137 and strontium-90 that pose great hazard if they get into the environment, where they can be absorbed by the bodies of humans and animals.

It is possible to "reprocess" spent fuel and extract the unexpended uranium-235 as well as the plutonium-239 that was created by the chain reaction. (After a year's operation a typical reactor core will have about 200 pounds of uranium and 500 pounds of plutonium.) The process involves a tedious, painstaking chemical separation of these fissionable elements from the waste. Once recovered, the uranium could be used to make more reactor fuel, and the plutonium could be used in a "breeder reactor" also to make more fuel. Despite the advantage of being able to eke out the supply of uranium, reprocessing has not been adopted in the United States partly because of concerns that the pure uranium and plutonium would be attractive to terrorists.

Managing Radioactive Waste

For every 1,000 megawatts of annual power production, a nuclear plant produces about 20 metric tons of spent fuel, or about 2,200 tons per year. By 2011 this amounted to more than 71,000 metric tons in the United States alone. This spent fuel is a form of what is known as high-level radioactive waste. It amounts to only about 0.1 percent of radioactive waste shipments, but it represents the greatest risk in case of accident. As a result of lack of agreement on how to proceed, much of the spent fuel in the United States has ended up being stored at the reactor site, immersed in a steel-lined pool where the water provides both cooling and radiation shielding, or in dry cask storage. In order, the states with the most of this waste in 2011 were Illinois (9,300 tons), Pennsylvania (6,400 tons), South Carolina (4,300 tons) and about 3,780 tons each in New York and North Carolina.

Another kind of nuclear waste is much less radioactive than spent fuel but also much bulkier. It is called low-level waste. Low-level wastes are materials that are weakly radioactive or have been contaminated with relatively small amounts of radioactive material. Radioactive materials are used in a variety of applications that lead to the generation of a stream of low-level nuclear waste. These include radioisotopes used in hospitals for diagnostic tests and cancer treatment as well as radioactive tracers used to test materials in factories or to analyze substances in laboratories. Some consumer products, such as illuminated watch dials and smoke detectors, contain small amounts of radioactive material. Other low-level waste includes gloves and protective clothing worn by medical or nuclear power plant workers. About two-thirds of low-level waste comes from the nuclear industry, including uranium mining and refining, fuel fabrication, and the operation of nuclear power plants. Use of techniques that minimize waste creation has reduced the amount of low-level waste by about half since 1980, but it still amounts to millions of cubic feet annually.

The Low Level Waste Policy Act of 1980 made each state responsible for the disposal of wastes generated within the state. Because one facility could handle wastes from several states, there has been an attempt to build such regional facilities. However, the familiar response of NIMBY ("not in my back-yard") means that the neighbors of any such proposed waste site will generally oppose it vociferously, for example, in 1993 California approved a low-level waste storage site at Ward Valley. But because that site was on federal land, Interior Secretary Bruce Babbitt was able to insist on an independent review by the National Research Council, together with the Bureau of Land Management and the U.S. Geological Survey. This complex environmental review and regulatory process makes it possible to bog down a proposed site almost indefi-nitely. Low-level waste sites are in operation in Barnwell, South Carolina; Beatty, Nevada; Richland, Washington; and Clive, Utah. In addition, the DOE manages dozens of other low-level waste sites, of which the largest are at government lab-oratories and nuclear weapons–related facilities.

In addition to the spent fuel from power plants and the remains of uranium mining and refining, there are also large quantities of waste and hundreds of contaminated sites result-ing from activities over more than half a century, particularly the production of thousands of nuclear weapons.

The Environmental Management (EM) office of the Department of Energy noted that in 2006 it was successful in doing the following:

> completing the cleanup at Fernald and Rocky Flats, along with these other sites: Columbus, Lawrence Livermore National Laboratory Main Site, and Kansas City Plant. In addition, despite numerous regulatory and technical

challenges, EM has made significant cleanup progress at its larger sites, such as:

- Stabilizing and packaging for disposition all plutonium residues and metals of oxides (Rocky Flats, Savannah River Site, Hanford);
- Producing well over 2,000 cans of vitrified high-level waste from highly radioactive liquid wastes (Savannah River Site, West Valley Demonstration Project);
- Retrieving and packaging for disposal over 2,100 metric tons of spent nuclear fuel from K-basins on the Hanford site to protect the Columbia River;
- Characterizing, certifying, and shipping close to 40,000 cubic meters of transuranic waste from numerous sites to the Waste Isolation Pilot Plant for permanent disposal;
- Disposing of close to one million cubic meters of legacy low-level waste and mixed low-level waste;
- Eliminating 11 of 13 high-risk material access areas through material consolidation and cleanup;
- Cleaning up the Melton Valley area at the Oak Ridge Reservation and completing the decontamination and decommissioning of three gaseous diffusion plants at the Oak Ridge Reservation (K-29, 31, 33); and
- Disposing of over 8,500 tons of scrap metal from Portsmouth, Ohio (Department of Energy 2007)

Search for a Permanent Solution

Notice that many of the accomplishments listed by the DOE involve "stabilizing," "retrieving and packaging," and "shipping." But where will these tons of waste finally end up? Low-level waste, such as that resulting from medical or industrial use of radioisotopes, while rather bulky, has been stored and managed in numerous facilities (typically a sort of pit or landfill) under NRC regulations. Such sites are expected to be stable and can be maintained indefinitely if properly monitored. Indeed much of the least radioactive low-level waste can simply be stored on site until its radiation drops below the regulatory limit, when it can be disposed of as ordinary waste, with due regard for any chemical hazards or biohazards.

High-level waste is a different matter. It includes plutonium reprocessed from decommissioned nuclear warheads and spent fuel from nuclear power plants. As a result of reactor operation, the nuclear fuel (usually U-235) reaches a point where it is no longer concentrated enough to efficiently sustain a reaction. The fission process also breaks uranium atoms into lighter radioactive isotopes such as strontium-90 and cesium-137. These are highly radioactive but have a relatively short half-life of about 30 years. The other major fission product, plutonium-239, is less radioactive but has a much longer half-life of 24,000 years. (Plutonium is also chemically toxic.) This mixture of different kinds of radiation emitters and the fact that spent fuel continues to generate considerable heat complicates the storage and disposal of these materials. For decades now high-level waste has been stored near nuclear power plants or other facilities either submerged in pools of water or in dry cask storage. (Waste is often vitrified, or turned into a glass-like substance that is less likely to react with the surrounding environment.)

In the 1990s the Nuclear Energy Institute estimated that 25 U.S. nuclear plants would run out of on-site storage space by the year 2000. Congress required that a temporary waste facility be ready by 1998, but in the Nuclear Waste Policy Act of 1982, the Department of Energy agreed to accept shipments of spent fuel from commercial nuclear reactors as of January 31, 1998.

Since the act was passed, utilities have charged their customers a penny for every kilowatt-hour generated by a nuclear reactor, creating a fund to be used for the development of a permanent waste disposal site. This fund has now grown to more than $10 billion. But finding a place to put the waste sites became a political nightmare.

After a number of proposals fell through, in 1987 Congress decided to focus all its waste disposal efforts on one site, Yucca Mountain, Nevada, about 100 miles northwest of Las Vegas. This huge site was to have more than 100 miles of underground tunnels about 1,000 feet below the surface, and 800 feet or more above the water table. But scientists would have to conduct a long painstaking testing program to try to prove that the stored waste could leak into the surrounding rock and potentially reach the water table, where it could get into the area's water sources.

Some scientists, such as two physicists at Los Alamos and a group at the Energy Department's Savannah River research facility, believe it is possible that thousands of years from now, when the metal waste containers have crumbled, the still very radioactive waste could pile together with rocks that will moderate the neutrons from the radioactive atoms, causing an explosive chain reaction. (Scientists have discovered that a rock formation in Africa functioned as a natural nuclear reactor in prehistoric times.) However, other scientists, including a review panel at the Los Alamos National Laboratory, say that the chance of this happening is remote, and the chance of a chain reaction within a fuel canister is also very low (Yucca Mountain Project, 1999).

By 1995, when it seemed clear facilities would not be ready in time, 19 utilities and 40 states and regulatory agencies filed suit, claiming that the federal government had reneged on its agreement after the utilities had paid millions of dollars.

The political battles continued in the new century. In 2002 the DOE recommended that the government proceed with the completion of Yucca Mountain as a long-term geologic repository for high-level nuclear waste. The governor of the state of Nevada objected, but Congress voted to go ahead and apply for a construction license from the NRC.

In May 2009 Energy Secretary Steven Chu told a technology writer, "Yucca Mountain as a repository is off the table." He noted that the NRC had found that "the dry cask storage at current sites would be safe for many decades." There was time to explore and develop better options, such as using high-energy "fast neutron reactors" to break down the long-lived waste. Waste that was unlikely to be of future economic value might be stored in natural salt domes that geologists believe are stable for 50 to 100 million years or more. Chu proposed that we "get some really wise heads together, and figure out how you want to deal with the interim and long term storage" (Bullis 2009).

In 2010 the DOE withdrew its NRC license application for Yucca Mountain. The project was for all practical purposes dead, although a number of states, concerned about temporary waste sites in their territory, filed legal objections that remained unresolved pending a final NRC decision. Altogether $32 billion had been collected from power companies to fund the project, of which $12 billion had been spent.

The delays in operating the Yucca Mountain site led some members of Congress to push for the building of a large aboveground temporary storage facility. However so far these proposals have also failed. A proposed storage facility in New Mexico was rejected by Mescalero Apache tribe members. In 2006 the Skull Valley Goshute Reservation in Nevada received an NRC licensed waste facility, but after a long, contentious struggle with Native Americans and anti-nuclear activists the company proposing the site voluntarily abandoned its license in 2013 without ever operating the facility. The piles of spent nuclear fuel (70,000 tons as of 2012) continue to grow at the rate of about 2,000 tons per year. Menawhile, the Fukushima disaster brought renewed concern in the United States about how an earthquake or other natural disaster might affect the nuclear waste now being stored at more than 120 nuclear plants in 31 states. Cliff Hamel, an economist and the chief author of a federal report recommending an interim waste site, nevertheless seems hopeful that a satisfactory solution to the waste problem will be found:

> "I think there is a change, a new openness to this as a solution," said economist Cliff Hamal, chief author of a federal report recommending the creation of an interim site. "California, Oregon, and other places that keep spent nuclear fuel are tired of having it. Centralizing this waste will lower the government's financial liability. The savings are substantial. And from a community's perspective, it will add well-paying jobs." (Davenport 2013)

Meanwhile, however, the dispersed waste facilities are potentially vulnerable to terrorists or just accidents. In early 2013, for example, six tanks of radioactive waste in Hanford, Washington (where plutonium had been processed for the Manhattan Project and later) were found to be leaking. It was uncertain when or whether the leaking waste might reach and contaminate groundwater or nearby rivers. Hanford stores more than 53 million gallons, of which as much of a million may have already leaked (Dininny 2013).

Transporting Nuclear Waste

Once a disposal site is established, the waste has to be shipped there safely. Each year about 3 million shipments of radioactive materials (mostly low level) are made throughout the United States. The nuclear industry points out that the safety record for these shipments is better than that for other forms of hazardous materials and that releases of radioactive materials to the environment are rare. Of these shipments only about 50 consist of spent reactor fuel. (It had been anticipated that when Yucca Mountain opened, between 400 and 500 casks of spent fuel would be shipped annually to the depository. With the cancellation, these casks will presumably remain in place.)

A typical shipment cask of high-level waste consists of spent fuel rods sealed in a stainless-steel cylinder and encased in heavy metal shielding and additional layers of steel. These casks are large: a cask shipped by truck is about 4 feet by diameter and 20 feet long, and it weighs about 25 tons. The larger casks used in rail transport are 8 feet in diameter and 25 feet long, and they weigh 125 tons. The NRC subjects these casks to brutal tests, including an 80 mile per hour (mph) crash into a thick concrete wall, a collision with an 80-mph locomotive, drops of up to 2,000 feet onto hard ground, and burning at 2,000 °F for more than an hour and a half. The actual shipping is regulated by the Department of Transportation (DOT) and regulators in the states through which the shipment will pass.

Especially since September 11, 2001, there has been considerable concern that nuclear waste shipments might either be sabotaged (to cause release of radioactive materials in the environment) or diverted so that the materials can be used to make nuclear weapons or (more likely) radiological "dirty bombs." As of 2010 regulations require armed escorts, continuous tracking of vehicles, and emergency procedures to respond to attempted attacks or theft. The shipment containers are also designed to be resistant to entry or tampering, and trucks have mechanisms that allow them to be remotely halted.

Decommissioning Nuclear Plants

The first generation of nuclear plants was designed to last 30 to 40 years. Newer plants are expected to last at least 60 years. Despite the successful efforts to extend the working lives of nuclear reactors, every nuclear power plant will reach the end of its useful life at some point. Decommissioning is the process of dismantling the plant and cleaning up any residual radioactivity in the area.

The NRC allows utilities to choose one of three alternatives for decommissioning an unwanted nuclear plant.

The DECON option involves hauling away the reactor vessel and radioactive components and decontaminating the area so that it can be used for any normal purpose. Because most of the power-generating machinery at a nuclear plant is not radioactive, it is possible to remove the reactor vessel and replace it with a conventional coal or gas boiler while reusing the turbines for power generation.

The choice to use the ENTOMB option means to permanently seal the reactor in concrete or other shielding material, which has been done with some government reactors but no commercial power plants. Finally, the SAFSTOR option can be chosen in which the reactor will be mothballed, allowing the radioactivity in the core to run down gradually under continuous supervision over a period of years. The plant can then be dismantled and decontaminated.

As of 2013 23 U.S. commercial reactors have been decommissioned, sometimes with fuel remaining stored on site. An interesting example is Rancho Seco in California, which suffered a failure of its steam generator control system that could, according to an NRC analysis, have led to a serious disaster. Long the focus of regional antinuclear protests and a series of voter referendums, California voters finally voted in 1989 to shut the plant down. In this case the SAFSTOR option was used, with final dismantling scheduled for 2018.

Decommissioning nuclear plants has turned out to be quite expensive. In the case of the Yankee Rowe and Maine Yankee plants where the DECON option was chosen, decommissioning costs exceeded $600 million per reactor, while the Connecticut Yankee plant's decommissioning cost $820 million. So far the most expensive decommissioning in the United States has been the Exelon Zion 1 and 2 reactor units, which is estimated to have cost as much as $1.1 billion. Following the Fukushima events, 11 Japanese reactors were badly damaged, and it is probable that many of them will eventually be decommissioned.

In general, decommissioning costs add to the full "life cycle cost" of a nuclear facility and help make nuclear power a relatively expensive energy option.

Nuclear Proliferation

A continuing international concern, heightened since the end of the Cold War in 1990, has been nuclear proliferation, or the unauthorized diversion of nuclear materials and related technology to countries or groups that currently lack such capabilities. In the first decade of the twenty-first century, such concerns focused particularly on Iraq (where they proved to be unfounded) and Iran (which claims it is enriching uranium for peaceful purposes). Another perennial issue is North Korea, which has conducted several nuclear tests and may have received considerable technical aid from Pakistan.

How is nuclear power related to nuclear proliferation? The fuel used in nuclear reactors (about a 3.5 percent concentration of uranium-235) is far too dilute to be used to make a nuclear bomb, which requires enrichment to a level of around 30 percent or higher. But it is certainly possible that a "rogue" nation or perhaps even an exceptionally well-financed terrorist group could, if it obtained commercial nuclear fuel, enrich the fuel using the same processes discovered in the 1940s during the Manhattan Project.

Further, high-level waste (spent fuel) might be clandestinely reprocessed to extract the uranium-235 or plutonium-239, either of which could be used to make a nuclear weapon. To reduce the proliferation of nuclear weapons to nonnuclear nations or terrorists, the United States has agreed to accept spent fuel from some 40 nations with nuclear research reactors, including Canada, Japan, and Australia.

One criticism of the expansion of nuclear power is that it would proportionally increase the risk of proliferation. As noted in a study of the effects of nuclear expansion:

large civilian nuclear energy programs can—and have—brought states quite a way towards developing nuclear weapons; and it has been market economics, more than any other force, that has kept most states from starting or completing these programs.

. . . if nuclear power evolves into the quickest and least expensive way to produce electricity while abating carbon emissions, little short of a nuclear explosion traceable to a "peaceful" nuclear facility is likely to stem this technology's further spread—no matter what its security risks might be. (Sokolski 2010)

On the other hand, nuclear power reactors might actually be used to reduce the supply of the highly enriched fuel needed to make a nuclear weapon. With the end of the Cold War, the world was suffering a plutonium glut. The dismantling of former U.S. and Soviet nuclear weapons is likely to yield 100 tons of plutonium, and the recovery of plutonium from spent reactor fuel amounts to about 20 tons per year. In its highly concentrated form, plutonium is compact and relatively easy to conceal and transport. This means that a terrorist group might obtain the few kilograms needed to create a "low-yield" but still devastating nuclear weapon, using plans that are now widely available.

One way to forestall this threat is to mix the plutonium with uranium into a "mixed oxide" fuel that could not be immediately used for nuclear weapons but could be used to generate electricity in standard nuclear reactors. Unfortunately, the cost of producing this fuel is currently about seven times that of conventional uranium fuel. Alternatively, the plutonium could be irradiated in a reactor and combined with fission products so that it would be very difficult for someone to process it for weapons use. Finally, plutonium that has been extracted from power reactor fuel could be "unreprocessed" by mixing it back into the spent fuel wastes from which it was extracted.

Existing high-level wastes containing substantial elements of plutonium or other heavy isotopes might also be "burnt" in this way to make the waste less attractive to would-be diverters.

Another proposal would use surplus military plutonium in a "triple play" reactor that would simultaneously deplete the plutonium, produce tritium (a radioactive gas needed for new nuclear warheads), and generate electricity. But this proposal is opposed by opponents of both nuclear power and the further development of nuclear weapons.

Nuclear Terrorism

As worrisome as nuclear "rogue states" might be, to many people the most frightening form of nuclear proliferation is the acquisition by a terrorist group such as Al Qaeda of sufficient nuclear material to make a nuclear or radiological weapon. The technical skills and facilities needed to enrich uranium or make even a crude nuclear weapon are probably beyond the capabilities of most terrorist groups. A more likely scenario might involve nuclear weapons being obtained as a result of the chaotic collapse of an existing nuclear power such as Pakistan, or possibly the clandestine transfer of such weapons from a government such as Iran to a militant group.

The kind of material that might be diverted from commercial nuclear plants or their waste would more likely be used to make a radiological weapon or so-called dirty bomb. This is basically a conventional explosive bomb packed with radioactive material that could be widely dispersed when the bomb is set off. Even if it did not directly kill or injure many people, such an attack could contaminate a wide area, forcing evacuation and a very expensive decontamination process. The psychological effects of such an attack could also be considerable, given the general public fear and lack of specific knowledge of radiation.

The International Atomic Energy Agency places a strong emphasis on "nuclear security" which it defines as:

The prevention and detection of, and response to, theft, sabotage, unauthorized access, illegal transfer or other malicious acts involving nuclear or other radioactive substances or their associated facilities. It should be noted that "nuclear security" includes "physical protection," as that term can be understood from consideration of the Physical Protection Objectives and Fundamental Principles . . .

Similarly to "safety culture," the IAEA defines "Nuclear Security Culture" as "The assembly of characteristics, attitudes and behavior of individuals, organizations and institutions which serves as a means to support and enhance nuclear security." (International Atomic Energy Agency 2008)

The International Convention for the Suppression of Acts of Nuclear Terrorism defines nuclear terrorism as follows:

1. Any person commits an offence within the meaning of this Convention if that person unlawfully and intentionally:
    (a) Possesses radioactive material or makes or possesses a device:
        (i) With the intent to cause death or serious bodily injury; or
        (ii) With the intent to cause substantial damage to property or to the environment;
    (b) Uses in any way radioactive material or a device, or uses or damages a nuclear facility in a manner which releases or risks the release of radioactive material:
        (i) With the intent to cause death or serious bodily injury; or
        (ii) With the intent to cause substantial damage to property or to the environment; or
        (iii) With the intent to compel a natural or legal person, an international organization or a State to

do or refrain from doing an act. (United Nations 2005)

In an attempt to make access to facilities by terrorists as difficult as possible, the IAEA has developed an approach called Design Basis Threat, or DBT. This involves a comprehensive approach to analyzing the design of an existing or planned nuclear facility as well as its operating procedures with a view to:

- Insider and/or external adversaries. A potential adversary is any individual or group of individuals deemed to have the intent and/or the capabilities to conduct a malicious act.
- Malicious acts leading to unacceptable consequences. Measures need to be taken to prevent malicious acts, such as unauthorized removal of material or sabotage, and protect against their unacceptable consequences.
- Attributes and characteristics. The attributes and characteristics of potential adversaries describe their capabilities to carry out a malicious act. Capabilities may include weapons, explosives, tools, transportation, insiders and insider collusion, skills, tactics, and number of individuals. These capabilities help determine the detection, delay, and response criteria for designing and evaluating an effective physical protection system.
- Design and evaluation. A DBT, which is defined at the state level, is a tool used to help establish performance requirements for the design of physical protection systems. A DBT is also used to help operators and state authorities assess the effectiveness of the systems to counter adversaries by evaluating the systems' performances against their capabilities described in the DBT. (International Atomic Energy Agency 2012)

The Non-Proliferation Treaty and Nuclear Security

Currently 190 nations are signatories to the Nuclear Non-Proliferation Treaty (NPT). This treaty took effect in 1970 and was extended indefinitely in 1995. The treaty outlines the following tasks:

> Undertaking to cooperate in facilitating the application of International Atomic Energy Agency safeguards on peaceful nuclear activities,
>
> Expressing their support for research, development and other efforts to further the application, within the framework of the International Atomic Energy Agency safeguards system, of the principle of safeguarding effectively the flow of source and special fissionable materials by use of instruments and other techniques at certain strategic points,
>
> Affirming the principle that the benefits of peaceful applications of nuclear technology, including any technological by-products which may be derived by nuclear-weapon States from the development of nuclear explosive devices, should be available for peaceful purposes to all Parties of the Treaty, whether nuclear-weapon or non-nuclear weapon States,
>
> Convinced that, in furtherance of this principle, all Parties to the Treaty are entitled to participate in the fullest possible exchange of scientific information for, and to contribute alone or in cooperation with other States to, the further development of the applications of atomic energy for peaceful purposes,
>
> Declaring their intention to achieve at the earliest possible date the cessation of the nuclear arms race and to undertake effective measures in the direction of nuclear disarmament. (International Atomic Energy Agency 1970)

In 2009 President Obama announced that at the Nuclear Security Summit that "we will advance our goal of securing all the world's vulnerable nuclear materials within four years" (Obama 2009). Secretary of State Hilary Clinton called the summit "an unprecedented gathering that will help promote a common understanding of the threat of nuclear terrorism and build international support for effective means of countering the threat"(Clinton 2009).

Leaders of 47 countries attended the 2010 summit, including many heads of state. Vice President Biden described the urgency of the problem: "We cannot wait for an act of nuclear terrorism before coming together to share best practices and raise security standards, and we will seek firm commitment from our partners to do just that" (Biden 2010).

In 2012 at the next meeting, President Obama offered this assessment of U.S. efforts:

Over the past two years, the questions have been different—would we back up our words with deeds; would we sustain our cooperation. Today, here in Seoul, we can answer with a resounding yes. We are fulfilling the commitments we made in Washington. We are improving security at our nuclear facilities. We are forging new partnerships. We are removing nuclear materials, and in some cases, getting rid of these materials entirely. And as a result, more of the world's nuclear materials will never fall into the hands of terrorists who would gladly use them against us. (Obama 2012)

However, the location of this meeting—Seoul, South Korea—was perhaps more significant than what was said there. That country's bellicose neighbor, North Korea, had already tested two nuclear weapons more or less successfully. By the end of the year, North Korea had finally succeeded in launching a satellite in orbit, demonstrating a capability that most Western observers agreed was really about developing an intercontinental ballistic missile (ICBM) that could eventually carry a nuclear warhead. And by April 2013, in an atmosphere of growing tensions, North Korea announced it would reactivate its plutonium production facility, which had been shut down as part of a 2007 agreement.

North Korea received much of its initial nuclear enrichment equipment from Pakistan. Current events in North Korea and (to a less advanced extent) Iran raise the question of whether antiproliferation agreements and monitoring can truly stop a determined nation from acquiring nuclear weapons. Is there a point at which military action might be used to destroy such capability before a nuclear arsenal can be built up and adapted for use in missiles? If not, would the sort of deterrence mechanisms that prevented the Cold War from becoming a nuclear war work in such cases? These questions remain very much open in the second decade of the twenty-first century.

The Search for Safety

Safety has evolved as an issue over the six decades that nuclear power plants have been operating. Awareness of the danger of radiation exposure led to concern about the safety of the growing number of nuclear power plants. As with many areas of controversy, early defenders of nuclear power tended to make earnest, reassuring statements.

Nuclear critics, however, point to a history of less-than-forthright statements by the nuclear establishment. Back in 1957, at the same time that it was assuring the public of the safety of the new nuclear power plants, the Atomic Energy Commission's Brookhaven National Laboratory issued a secret report that assessed the potentially devastating effects of a nuclear accident that might kill thousands of people, cause $7 billion worth of damage, and poison 50,000 square miles of land. And in 1977 a researcher at the Union of Concerned Scientists discovered the "nugget file," an internal file kept by an official at the Nuclear Regulatory Commission (NRC) that lists hundreds of instances of potentially serious equipment failures and operator errors.

Nuclear plant safety involves both the reactor core and the surrounding cooling systems. During the 1950s builders of the early reactors were most concerned with keeping the nuclear reaction under control, poised at the level of neutron production where each fission event led to an average of one additional event, resulting in steady power output. Such control can be difficult because conditions of temperature and pressure and the presence of fission products are constantly changing the absorption of neutrons and thus the rate of fission. However, an emergency system monitors these conditions, and if the reactor is in danger of going "supercritical," a safety device automatically inserts neutron-absorbing control rods into the core, halting the chain reaction. (As an additional measure, a neutron-absorbing liquid such as boric acid can also be injected into the core.) This controlled rapid shutdown event is called a scram.

The importance of the cooling system arises because the radioactive decay of uranium and other radioactive atoms in the core cannot be shut off. Even when a fission reaction is no longer possible, heat will still build up in the core because of this inherent radioactivity. As long as coolant (water, in most reactors) continues to circulate through the core, the heat is carried away and there is no problem. But if the supply of coolant is cut off, such as by a pipe break, a pump failure, or loss of electric power to the pump, some backup cooling or power system must take over, or a "loss of coolant accident" (LOCA) will occur.

To try to prevent such an accident, reactors have an emergency core cooling system (ECCS) that supplements that main system of cooling and heat exchange. This system can include pumps that can inject coolant into the core even while it remains pressurized. There can also be a system that depressurizes the reactor vessel by venting steam into the bottom of a pool of water. The reduced pressure makes it easier to get more coolant into the core.

If these systems fail (perhaps because of a power outage) or prove inadequate, the reactor will continue to overheat. As the core heats, it turns whatever water remains in the core to steam, dissipating it. As the core itself becomes exposed to the air, it heats rapidly, and the fuel itself may start to melt. This is the infamous "meltdown" that can potentially result in the molten fuel rupturing the reactor vessel and even the containment building. The worst-case result is a massive release of highly radioactive liquid or steam into the environment, possibly accompanied by an explosion or fire. (A reactor cannot explode like a nuclear weapon, as the latter requires a very precise formation of a supercritical mass.)

The final layer of defense is a series of physical containments. These include the cladding or coating around the fuel itself, the reactor vessel, and the containment building around the whole reactor assembly. If all else fails, a "core catcher" made of metal at the bottom of the reactor could dilute the melting core and make it easier to finally cool.

In 1963 researchers began to build the LOFT (Loss of Fluid Test Reactor) to determine exactly what takes place after coolant is lost in a reactor core. However, the project fell victim to bureaucratic delays and was never used for its intended purpose. Dr. Carroll Wilson, the first general manager of the Atomic Energy Commission (AEC) during the 1950s, later believed that "one of the grievous errors of the Atomic Energy Commission in the 1960s was the failure to carry out the repeatedly recommended experiment of running a [large-scale] reactor to destruction to find out what would really happen instead of depending on studies based on computer models for such vital information" (Norton 1982, 21).

As a result, the main source of data about what actually happens in a major reactor failure is from events such as Three Mile Island, Chernobyl, and now, Fukushima. After each event there are attempts to identify problems and develop new safety features, but also concerns about whether lessons have been sufficiently learned.

As the authors of a recent study in the *Bulletin of the Atomic Scientists* note, there are three goals in trying to make reactors safer: "to prevent serious consequences resulting from an accident (preventing the melting of reactor fuel if reactor cooling fails catastrophically); to mitigate damage by allowing for timely, economic restoration to normal operations ... and to delay potential serious impacts by providing ... sufficient warning time for the plant operator to prevent eventual serious consequences" (Rosner, Lordan, and Goldberg 2011).

The Union of Concerned Scientists, which has had been tracking ongoing safety issues at U.S. nuclear plants, "has found that leakage of radioactive materials is a pervasive problem at almost 90 percent of all reactors, as are issues that posea risk of accidents. (Cooper 2012)" Energy economist Paul Cooper notesthat one or more of the three issues that surfaced at Fukushima—seismic risk, fire hazard, and elevated spent fuel storage—are also faced by 80 percent of nuclear plants in the United States (Cooper 2012).

Does "Safety Culture" Work?

The ability to learn lessons and respond to new challenges is part of the human dimension of nuclear safety. There have been a number of efforts to inculcate what is called a safety culture in the nuclear industry. One definition of nuclear safety culture is "An organization's values and behaviors—modeled by its leaders and internalized by its members—that serve to make nuclear safety the overriding priority" (Institute of Nuclear Power Operators 2004, 3).

The Institute of Nuclear Power Operators points out the importance of "A safety-conscious work environment (freedom to raise concerns without fear of retribution)" (Institute of Nuclear Power Operators 2004, iii). Also, "Leaders recognize that production goals, if not properly communicated, can send mixed signals on the importance of nuclear safety" (Institute of Nuclear Power Operators 2004, 3). Also, "Nuclear technology is recognized as special and unique" (Institute of Nuclear Power Operators 2004, 6).

The question of how strong and effective safety culture is in the real world is an open one. One potentially serious incident that set off alarm bells both within and outside the industry occurred at the Davis-Besse nuclear power station in Oak Harbor, Ohio. In March 2002 workers at the plant discovered that the borated water coolant had leaked through cracked control rod mechanisms at the top of the reactor. The corrosive coolant had leaked through more than 6 inches (150 millimeters) of the reactor pressure vessel head. At the time the damage was discovered, only a thin (3/8-inch or 9.5-millimeter) layer of stainless steel cladding was holding back the high-pressure reactor coolant. If the coolant had burst through this last barrier, it would have rapidly drained, probably leading to a major loss of cooling accident.

After this problem was identified, a number of other safety issues were identified with various pumps and valves as well as with the plant's emergency generator and its electrical system. It cost the reactor operator $600 million to fix the problems and install safety upgrades, and the NRC imposed a fine of more than $5 million—the largest ever such penalty.

On top of this, criminal charges were made that some company engineers had overstated the thoroughness of safety inspections prior to the incident, as it was clear from the physical evidence that the corrosion must have taken place over a considerable time. This prosecution resulted in 2006 in what amounted to a plea bargain involving another $23.7 million in fines and $4.3 million in remediation. Several engineers were indicted, with mixed verdicts.

An evaluation of safety culture at the Davis-Besse power station concluded, among other things, that "An integrated and cohesive organizational safety leadership process does not yet exist" and "Efforts to improve future performance by learning from the Station's past performance, from others' performance, and from the day-to-day implementation of the organization's programs and processes are not systematic or recognized to be of high value for the organization" (Performance, Safety, and Health Associates 2003, iv).

According to Charles Ferguson, "a safety culture has succeeded best in work environments in which potential hazards are discussed openly and plant personnel are encouraged to bring attention to potential problems early, without fearing repercussions. In fact, the employees should be rewarded for drawing attention to safety concerns. While at times personnel have been reprimanded for doing such, whistleblower protection laws are needed to assure workers. The highest performing and safest plants tend to foster openness" (Ferguson 2011a).

Coping with Natural Disasters

The Fukushima earthquake and tsunami of March 2011 and the resulting multiple reactor meltdowns forcefully brought a new factor into the nuclear safety equation. Reactors are designed to deal with a normal environment, including storms and even earthquakes if they are to be expected in the area. But what happens if an unusually severe event—or, as in this case, a combination of severe events—occurs?

Has the danger of flooding and its effects on electrical systems been adequately addressed? According to the Union of Concerned Scientists:

> For several years prior to the March 2011 tsunami that crippled the Fukushima nuclear plant in Japan, the NRC had known that 35 U.S. reactors ... faced flooding hazards potentially greater than they are designed to withstand. Nuclear plants are built next to rivers, lakes, and oceans because they require vast quantities of cooling water to carry away the large amounts of waste heat. (Today's nuclear plants produce two units of waste heat for every unit of electricity they generate.) Many nuclear plants along a river have one or more dams located up that

river. If a dam failed, the ensuing flood waters could over-whelm the plant's protective barriers just as the tsunami crashed over Fukushima's seawall. (Union of Concerned Scientists 2012)

The NRC is in the process of re-evaluating the risks and con-sequences of such flooding, whether caused by a dam failure or a natural event.

Advocates of nuclear power draw a different lesson from Fukushima. They note that the Onagawa nuclear power plant, a facility that was actually closer to the epicenter of the earth-quake than Fukushima and that was also flooded by a tsunami, withstood both quake and tsunami without incident. The plant was surrounded by a 46-foot (14-meter) seawall that prevented flooding. The plant was shut down successfully as required by procedure, and a fire in a turbine building was extinguished. Remarkably, several hundred people who had lost their homes in the flooding took refuge temporarily at a gymnasium at the plant.

According to Reuters reporter Risa Maeda, "Onagawa may now serve as a trump card for the nuclear lobby—an example that it is possible for nuclear facilities to withstand even the greatest shocks and to retain public trust" (Maeda 2011).

A multiyear, multimillion-dollar severe accident analysis (State-of-the-Art Reactor Consequence Analyses [SOARCA]) conducted by the Nuclear Regulatory Commission and released on February 1, 2012, found that in the event of a total loss of electrical power and the failure of all permanent safety systems (similar to the accident in Japan), the public would be protected effectively with no significant near-term or long-term health effects. The report concludes that

in all scenarios studied, even the unmitigated ones, events progress more slowly and release much less radioactive material than an extremely conservative 1982 siting study had found. Among its conclusions: "the calculated risks of public health consequences from severe accidents . . . are very small." (Nuclear Regulatory Commission 2012)

Richard Lester of MIT cautions against being too quick to find reasons why the particular circumstances of the disasters at Chernobyl and Fukushima were so different that such an event could not happen in the United States. He notes:

Another lesson from Fukushima is that failures of nuclear governance can happen anywhere. It's worth noting that the three most serious breakdowns of this type in history occurred in vastly different political and institutional settings: the United States, the former Soviet Union, and now, Japan. It would be unwise for nuclear leaders in any country to assume that it could never happen to them. Indeed, we might reasonably conclude from the evidence that there is no national system within which major failure cannot occur.

There are some universal principles of effective nuclear governance that apply everywhere regardless of national context. Principles like the transparency of decision-making, independence of regulation, and the importance of maintaining a safety culture in nuclear organizations. (Lester 2013)

New Reactor Designs

In April 2013 Gregory B. Jaczko, a former chairman of the Nuclear Regulatory Commission, made a remarkable statement. Looking at the wake of the Fukushima disaster, Jaczko reflected:

I was just thinking about these [safety] issues more, and watching as the industry and the regulators and the whole nuclear safety community tries to address these very, very difficult problems. Continuing to put Band-Aid on Band-Aid is not going to fix the problem. (Wald 2013)

The particular problem Jaczko had in mind arose from an inherent characteristic of all current-generation commercial nuclear power plants. Since nuclear fuel is intensely radioactive, it is constantly generating a lot of heat even when the reactor is not being operated to produce power. If there is a loss of cooling (as at Fukushima), even shutting down the chain reaction will not stop the "decay heat" from building up in the core. If this goes on long enough, the core will melt down—likely with disastrous results.

But what if the reactor fuel were encapsulated in such a way that heat would dissipate even without a cooling system? Or suppose the reactor were small enough that the fuel simply couldn't get hot enough to melt? Jaczko suggested that the existing large reactors be replaced as soon as possible with smaller ones—and that, meanwhile, the industry no longer try to extend the lives of existing plants.

Industry spokespersons disagreed with Jaczko's sweeping conclusions. As a practical matter, replacing or phasing out the nation's 104 nuclear power reactors would be a massive and extremely costly undertaking.

However, new reactor concepts and designs, while still largely confined to the drawing board, do offer "meltdown-proof" designs among other features that could bring improved safety. To understand how these new reactors would differ from current ones, let us first put them in the context of the evolution of reactor design. Nuclear reactors have been classified into four generations.

Generation I represents the early prototype or experimental reactors, such as Shippingport. Currently all 104 operating nuclear power reactors in the United States are classified as Generation II. These were originally designed to have a working life of about 30 years, but many have been modernized with improved safety systems and extended to a 60-year design life.

Generation III basically represents substantial improvements on Generation II but not a radical redesign, featuring improved efficiency (less fuel needed per unit of power). Improved safety features reduce the chance of core damage in an accident, and the design life may be extended to as much as 80 years. Most new reactor designs currently proposed are classified as Generation III or III+.

Generation IV represents designs that are currently theoretical and speculative. Some of these are designed to run at very high temperatures using coolants such as helium or molten salt. These "fast" reactors produce neutrons energetic enough to "burn" waste and create more fuel. Some Generation IV design features can be found in pebble bed modular reactors (PBMRs) that encapsulate fuel in ceramic coated "pebbles."

Besides attempting to offer greater efficiency (and thus more hopeful economics), new reactor designs have claimed a variety of safety features. This acknowledges that there are limits to how safe current reactors can be made to be. Although more backup systems such as pumps and generators can be added, such precautions increase both the cost and the complexity of the system.

The basic approaches to improving safety involve making the reactor more robust, that is, able to withstand not just "reasonably" expected stresses but also "unreasonable" stress that might be inflicted by an event such as a major earthquake or the impact of a hijacked aircraft. The second way to improve safety is by having multiple safety and backup systems, such as backup power and multiple sources of coolant.

There are other design features that could improve safety. For example, by building a lower-power (600 megawatts instead of 1,200 megawatts) reactor, the heat is reduced to the point where it can be handled by a simple system where water flows in through gravity and circulates by convection (warm water rising because it is lighter). The nice thing about gravity is that you don't have to worry about it failing. The fuel core itself can also be redesigned so that the fuel pellets are made smaller and are encapsulated by ceramic materials that can withstand extremely high temperatures. The fuel would be arranged in such a way that it would not melt even after a total loss of coolant. Critics, however, have argued that these new designs may not be as safe as claimed and that adopting them might create a level of overconfidence that would lead to a slackening of regulation, oversight, and monitoring.

Small Modular Reactors

A new alternative design called the small modular reactor (SMR) may offer a number of advantages. Today's commercial reactors are large-scale units. Although built using accepted designs and components, each installation is built in place, something like a house being built to an architect's plans. SMRs are intended to be completely assembled at the factory (taking advantage of more efficient production techniques) and then brought to the site where they are to be used.

Paul G. Lorenzini, chief executive officer (CEO) of NuScale Power, argues that SMRs offer several advantages: "They significantly reduce the demand for front-end capital outlays; they offer enhanced safety, critical post-Fukushima; and they reach markets which have, until now, been inaccessible to nuclear power because there were not suitable to large plants" (quoted in Wright 2012).

David Szondy notes that the much lower construction and operating costs make them "very attractive to poorer, energy-starved countries; small, growing communities that don't require a full-scale plant; and remote locations such as mines and desalination plants." SMRs are also more versatile. Instead of requiring a nearby water source, they "can be cooled by air, gas, low–melting point metals, or salt" (Szondy 2012).

However, Peter A. Bradford of the Union of Concerned Scientists notes, "No SMRs are licensed or ordered. We have no idea what they will cost. They have no likelihood of being competitive with natural gas for many years if current price projections are accurate" (quoted in Wright 2012).

SMRs may make it possible to provide electric power in remote areas where conventional power plants would not be economically practicable. However, in such areas there are likely to be fewer trained personnel available, which would increase the need for simplicity of operation and the use of passive and redundant safety features that would not require human input. Edwin Lyman, senior staff scientists with the Union of Concerned Scientists, expressed doubts that adequate staffing could be found: "Nuclear power requires high-level security and expertise to operate safely. It seems like something that should be concentrated rather than distributed [dispersed]" (quoted in Smith 2010).

A number of companies are developing SMRs and have begun preliminary discussions with the NRC about modifying licensing procedures to accommodate them. These include Babcock & Wilcox (mPower reactor), GE-Hitachi (PRISM reactor), Hyperion Power Generation (HPM reactor), Westinghouse (IRIS reactor), as well as NuScale Power and Toshiba for as yet unnamed reactors. However, adapting the cumbersome NRC licensing process to evaluate these new small reactors may prove difficult. Normally the NRC first evaluates and certifies the reactor design and then uses separate procedures for the actual siting and operating of the reactor. The NRC has been trying to determine how these procedures can be modified and perhaps streamlined by applying existing experience with small research reactors that are perhaps more comparable to the proposed SMRs.

"Fast" and Breeder Reactors

As far as the fission process is concerned, most SMRs are "fast" reactors in which neutrons are not slowed down by a moderator. To get sustainable fission with fast neutrons requires a different mix or type of fuel, perhaps plutonium instead of uranium-235. Fast reactors can also create new fuel through absorption of neutrons, thus functioning as a breeder reactor. Such reactors can be designed to require less frequent refueling.

According to Stefano Monti, team leader for the IAEA's Fast Reactor Technology Section in the Department of Nuclear Energy:

> [F]ast reactors operating in a closed fuel cycle would be able to provide energy for thousands of years as well as [ease] concerns about waste. The fast breeder technology has the potential to make the production of energy from uranium 100 times more efficient than with the existing thermal reactor, reducing the amount and toxicity of radioactive waste, as well as the heat emanating from the waste, and also shortening the waste's hazardous lifetime span. (quoted in Rickwood and Kaiser 2013)

The "closed fuel cycle" means that spent fuel is reprocessed rather than being disposed of or stored as waste. This can be done because spent fuel contains material that is "fertile," that is, while not fissionable in its current form, it can be converted to other fissionable isotopes.

In March 2013 the IAEA hosted a conference in Paris to which more than 600 experts came "to review fast reactor and fuel cycle technology advances, safety, [and] economic and proliferation-resistant issues" (Rickwood and Kaiser 2013).

However, nuclear critics suggest that this technology would create supplies of enriched nuclear fuel that would be attractive to terrorists and that the handling of such materials would also raise serious environmental concerns and risks.

Fusion?

In 2015, according to the movie *Back to the Future*, Dr. Emmett Brown no longer has to mess around with stealing dangerous plutonium from equally dangerous terrorists.

Just pop the lid of the coffee maker–sized Mr. Fusion Home Energy Reactor and dump in any old garbage (such as a banana peel, some beer, and then the beer can). Evidently "Mr. Fusion" can convert any sort of matter into enough energy to run a DeLorean time machine.

Of course, as this text is written it's almost 2015 and while we have shrunk powerful computers down to pocket size and may soon have cars that drive themselves, progress toward viable fusion power has been much slower. As French physicist and Nobel laureate Pierre-Gilles de Gennes has remarked, "We say that we will put the sun into a box. The idea is pretty. The problem is, we don't know how to make the box" (quoted in Kaku 2008, 47).

Triggering a fusion event is not the basic problem in fusion power—after all, the hydrogen bomb does that on a spectacular scale. Instead, the challenge is how to sustain a fusion reaction for more than a moment and how to get more energy out than one is putting in—a necessity for a practical power plant. There are two basic approaches. One is the use of powerful magnetic fields to contain the extremely energetic plasma (cloud of charged particles) in a donut-shaped ring called a tokamak. Invented by Soviet physicists in the 1950s, the world's most powerful tokamak device today is the Joint European Torus (JET). In 1997 the JET produced a peak power of 16.1 megawatts and sustained 10.1 megawatts of fusion power for about half a second. (Unfortunately, the power output was only 65 percent of the power needed to generate the magnetic field.)

Starting in 2007 the ITER (originally standing for International Thermonuclear Experimental Reactor) has been under construction in southern France. Costing about $14 billion and scheduled to become operational in 2020, the ITER is expected to, for the first time, produce more power than is put in—about 10 times more. It is estimated that ITER will generate about 500 megawatts of fusion power and sustain the reaction for up to 1,000 seconds (a bit over 16 minutes). If successful, ITER may show a viable path to commercial fusion power. On the other hand, the project has been plagued by cost overruns and engineering problems.

The other main approach to fusion power is called inertial confinement (ICF). This uses beams from a very powerful laser to rapidly compress and heat a pellet of deuterium-tritium fusion fuel. This has the same effect as the X-rays produced by the fission part of a thermonuclear bomb. The outer portion of the fuel pellet explodes into a cloud of plasma, which in turn forces the remaining fuel together. Shock waves form, increasing the density still more, which creates high-energy alpha particles whose energy then turns into heat. Under the proper conditions, a self-sustaining reaction called "ignition" can occur. (An additional complication is that a cascade of particular reactions must be managed so that enough tritium is being produced to be fed back into the process.)

A modest milestone was achieved in September 2013 at the National Ignition Facility (NIF) at the Lawrence Livermore Laboratory in California. The energy measured coming from the fusion in an energy pellet was greater than the amount of energy being absorbed by the fuel. However this was well short of true ignition which requires an energy output greater than the total energy coming from the 192 laser beams converging on the pellet.

However, while this experiment "has been described as the single most meaningful step for fusion in recent years" (Rincon 2013) three goals have yet to have been met: the ability to create a process that can be sustained indefinitely rather than in very short bursts; obtaining substantially more energy output than input; and developing an actual mechanism for collecting the output in the form of useful energy.

Although antinuclear activists have focused on today's fission rather than tomorrow's possible fusion power, they have made some criticism of efforts to develop viable fusion reactors. Criticisms include some safety issues such as contamination from radioactive tritium that is part of the fuel, as well as the unknown effects of the neutrons produced by sustained operation on the integrity of reactor confinement. However, the main criticism is that fusion is unlikely to become available in time to address climate change and represents a diversion of funds that would be better spent on proven renewable energy technologies such as solar and wind.

Fusion advocates counter that fusion produces hundreds of times less radioactive waste than equivalent fission plants and note that it is all short-lived waste. Further, fusion is safer because if the plasma were to escape (or the power shut off to lasers), the reaction would immediately stop. This is unlike the case with fission reactors, where the inherent radioactive heat of the fuel must be constantly cooled even when the reactor is not operating. Finally, the kind of fusion setup found in a power plant is completely different from that in a nuclear weapon, so there is no way to set off a thermonuclear explosion or divert material to make bombs.

Public Opinion Changing?

In democratic countries at least, it is the people who will eventually decide whether and how to develop nuclear power. The history of this technology has had many swings in popular opinion, ranging from early enthusiasm in the 1950s to second thoughts in the 1970s and 1980s punctuated by the disasters at Three Mile Island and Chernobyl as well as the growth of a large antinuclear movement.

Following the 2011 Fukushima disaster public support for building more nuclear power plants fell sharply. According to a CBS News poll, only 43 percent approved of building new plants, compared to the previous low of 46 percent in 1979 shortly after the Three Mile Island accident. The 2011 poll did find that more than two in three respondents still believed that American nuclear plants were safe. Sixty-two percent would oppose building a new plant in their community, and 31 percent were "very concerned" about a nuclear accident in the United States (65 percent were "somewhat concerned") (Montopoli 2011).

In a mid-2011 BBC poll that surveyed people in 23 countries, only 22 percent agreed that "nuclear power is relatively safe and an important source of electricity, and we should built more nuclear power plants." Seventy-one percent said that their country "could almost entirely replace coal and nuclear energy within 20 years by becoming highly energy-efficient and focusing on generating energy from the Sun and wind." Thirty-nine percent want to freeze the number of reactors in their country, while 30 percent want to shut down the existing reactors (Black 2011).

In general, the increase in opposition to expanding nuclear power was high in Germany, France, and Russia. It actually rose in Great Britain by 4 percent, but remained essentially unchanged in the United States at around 40 percent, as in China and Pakistan (Black 2011).

A Stalled Industry?

With no nuclear plants built for three decades in the United States, it has often appeared as though growth in the American nuclear industry had stalled if not reversed. One reason is obviously the decline in public confidence in nuclear energy in many if not all countries.

As the antinuclear reaction fueled by Three Mile Island continued, a growing number of nuclear plants that had been in the news began to shut down, including San Onofre unit 1, Rancho Seco (shut down by voter referendum), and Humboldt Bay in California; Trojan in Oregon; Yankee Rowe in Massachusetts; and Shippingport in Pennsylvania, which had been the very first plant built. The Shoreham nuclear power station in Long Island had just completed low-power testing when opposition by local and state officials forced it to shut down.

Rising costs and the changing economics of energy in the late 1970s and 1980s also played an important part in nuclear energy's apparently dismal future. Several factors led to nuclear power costs increasing rapidly in the 1970s and early 1980s. The nuclear industry stopped receiving government subsidies (except with regard to liability), and new regulations imposed significant costs.

As noted earlier, new supplies and lower prices for natural gas (the cleanest fossil fuel) may also be making nuclear power less attractive, at least for the near future. And since nuclear plants are so expensive, investors require near certainty that they will be cost competitive when they come online, which can be 10 or 15 years in the future.

Meanwhile, as the existing fleet of nuclear plants age, equipment failure begins to take an expensive toll. In 1986 a corroded pipe carrying superheated water in one reactor burst open and scalded four workers to death. The NRC imposed new regulations requiring close monitoring of all pipes and replacement of those showing corrosion or brittleness. The heat exchangers, which transfer heat from the core of the reactor to generate steam for the turbine, are also prone to cracking and leaking, sometimes releasing radioactive steam or water into the environment. The leaks are detected by monitoring systems, which minimizes such releases, but following a release, the plant must be shut down and expensive replacements made. For example, Baltimore Gas and Electric plans to spend $300 million to replace its steam generator as part of its license renewal, whereas Portland [Oregon] General Electric decided to close its Trojan plant rather than pay for expensive replacements. Even the reactor core itself becomes brittle after years of neutron bombardment, reducing its ability to contain melting fuel after a loss of coolant accident.

In the early 2000s a "nuclear renaissance" seemed possible, though, and it is possible that nuclear power may still be called upon to play a significant role in our energy future. Even though the specter of radioactive risk still haunts much of the public awareness of the nuclear industry, another specter has come to trouble our view of the future—carbon dioxide (CO_2) and other greenhouse gases.

Nuclear Power and Climate Change

Despite the serious economic and safety challenges to the industry, there have been two strong arguments for the expansion of nuclear power. The original argument had to do with making the United States less dependent on foreign oil, particularly supplies to the United States that could be cut off by hostile leaders or political instability.

In the early 1970s the Arab oil embargo and related events caused a spike in the price of oil on the world market. The results were painful prices and long lines at the gas pump, and what became known as "the energy crisis." Just before this happened, President Richard Nixon had launched a program he called Project Independence that proposed to expand use of nuclear power to reduce dependence on foreign oil. Despite sometimes considerable domestic opposition, countries such as Germany and France also planned to expand their use of nuclear power.

By the twenty-first century, however, oil prices, though higher, seemed to be roughly stable, coal was readily available (if environmentally problematic), and natural gas seemed to be getting cheaper. The prospect of accessing new domestic supplies of gas and oil (primarily through hydraulic fracturing, or "fracking"), together with ongoing improvements in fuel efficiency and conservation, has made the concern about energy independence recede somewhat. At the same time these factors have also reduced the incentive to build capital intensive, potentially risky nuclear plants.

Today widespread and increasingly urgent concern about global warming probably offers the best hope for advocates of a new expansion of nuclear power in the twenty-first century. Emissions (especially carbon dioxide) from burning fossil fuels trap heat in the earth's atmosphere and seem to be leading to a rise in the average world temperature that could have a devastating effect on coastal communities, agriculture, and other areas. In response to this concern, the Kyoto Protocol of 1992, put forth a goal of reducing greenhouse emissions to below 1990 levels early in the twenty-first century. But how can this be done while still providing the energy needs of both the developed world and the emerging economies of nations such as China and India?

Of the non–fossil fuel power sources, water power, once considered an attractive alternative, seems to have little potential for further growth. Hydroelectric dams have already been built in the most likely places, and many people oppose new ones because they disrupt the environment. Solar power offers clean energy and useful heating, but it lacks the concentration to feed the needs of cities and large factories. Nuclear fusion power (created by fusing, or combining, light atoms) would release far less radiation than nuclear fission, but after decades of research it has yet to become commercially viable.

For its advocates, only nuclear fission power seems to offer the combination of concentrated energy and a lack of conventional pollution and greenhouse gases. Famed environmental scientist James Lovelock offered such an opinion in a 2005 magazine article:

> The benefits of using nuclear energy instead of fossil fuels are overwhelming. We know nuclear energy is safe, clean and effective because right now, 137 nuclear reactors are generating more than one-third of Western Europe's electricity and 440 in all are supplying one-seventh of the world's. . . . To phase out nuclear energy just when we need it most to combat global warming is madness. (Lovelock 2005, 5)

Nuclear advocates go further, though. They believe that not only should we not phase nuclear out of the energy mix, but only nuclear can meet the continuing and growing demand for electricity in the newly industrializing world without adding to the climate problem. Charles Ebinger and John Banksof the Brookings Institution note that:

> With global electricity demand increasing dramatically and greenhouse gas emissions and energy security becoming national priorities, developed and developing countries alike are reexamining nuclear energy as a means of providing a reliable and scalable source of low-carbon power. (Ebinger and Banks 2011, 1)

However, other energy experts looking at the options for dealing with the need to reduce carbon emissions sum up the major objections to nuclear power as reasons why it is not getting more support even in the face of a perceived crisis:

> In spit of [its negligible greenhouse gas emissions], nuclear power is not seen as the solution to the global warming problem in many countries. The main issues are (1) the high costs compared to alternative CCGTs [Combined Cycle Gas Technologies], (2) public acceptance involving operating safety and waste, (3) safety of radioactive waste management and recycling of nuclear fuel, (4) the risks of nuclear fuel transportation, and (5) nuclear weapons proliferation. (Metz 2011, 65)

Former vice president (and presidential candidate) Al Gore, who has been a prominent advocate for dealing with climate change, in a 2006 speech summarized his concerns with expanding nuclear power:

> Many believe that a responsible approach to sharply reducing global warming pollution would involve a significant increase in the use of nuclear power plants as a substitute for coal-fired generators. While I am not

opposed to nuclear power and expect to see some modest increased use of nuclear reactors, I doubt that they will play a significant role in most countries as a new source of electricity. The main reason for my skepticism about nuclear power playing a much larger role in the world's energy future is not the problem of waste disposal or the danger of reactor operator error, or the vulnerability to terrorist attack. Let's assume for the moment that all three of these problems can be solved. That still leaves two serious issues that are more difficult constraints. The first is economics; the current generation of reactors is expensive, take a long time to build, and only come in one size— extra large. In a time of great uncertainty over energy prices, utilities must count on great uncertainty in electricity demand—and that uncertainty causes them to strongly prefer smaller incremental additions to their generating capacity that are each less expensive and quicker to build than are large 1000 megawatt light water reactors. Newer, more scalable and affordable reactor designs may eventually become available, but not soon. Secondly, if the world as a whole chose nuclear power as the option of choice to replace coal-fired generating plants, we would face a dramatic increase in the likelihood of nuclear weapons proliferation. During my 8 years in the White House, every nuclear weapons proliferation issue we dealt with was connected to a nuclear reactor program. Today, the dangerous weapons programs in both Iran and North Korea are linked to their civilian reactor programs. Moreover, proposals to separate the ownership of reactors from the ownership of the fuel supply process have met with stiff resistance from developing countries who want reactors. As a result of all these problems, I believe that nuclear reactors will only play a limited role. (Gore 2006)

The Union of Concerned Scientists, often critical of the nuclear industry, suggests that the risk of catastrophic accidents or terrorist attacks would also increase if more nuclear plants were to be built:

> UCS recognizes the need for a fresh examination of all possible options for coping with climate change, but it must be borne in mind that a large-scale expansion of nuclear power in the United States or worldwide under existing conditions would be accompanied by an increased risk of catastrophic events—a risk not associated with any of the non-nuclear means for reducing global warming.
>
> These catastrophic events include a massive release of radiation due to a power plant meltdown or terrorist attack, or the death of tens of thousands due to the detonation of a nuclear weapon made with materials obtained from a civilian—most likely non-U.S. nuclear power system. Expansion of nuclear power would also produce large amounts of radioactive waste that would pose a serious hazard as long as there remain no facilities for safe long-term disposal. (Union of Concerned Scientists 2007)

Much of the debate includes the tacit assumption that nuclear power in fact could be successfully expanded if it were decided this was the best way to address the climate crisis. However, a number of other experts have raised questions about this assumption as well.

Physicist and engineering professor Derek Abbot has suggested that there are fundamental limits to how much nuclear power can be expanded. Besides the question of fuel, the ultimate bottleneck might be

> an increased demand for rare metals—such as hafnium, beryllium, zirconium, and niobium, for example—[that] would also increase their price volatility and limit their rate of uptake in nuclear power stations. Metals used in the reactor vessel eventually become radioactive and, on decommissioning, those with long half-lives cannot be

recycled on timescales useful to human civilization. Thus, a large-scale expansion of nuclear power would reduce "elemental diversity" by depleting the world's supply of some elements and making them unavailable to future generations. (Abbot 2012, 68)

Another widely held assumption is that nuclear power is in fact carbon free. Actually, most experts would agree that the nuclear fuel cycle as a whole does have a carbon footprint. Prominent nuclear critic Dr. Helen Caldicott notes that nuclear power is not really carbon free:

large amounts of traditional fossil fuels are required to mine and refine the uranium needed to run nuclear power reactors, to construct the massive concrete reactor buildings, and to transport and store the toxic radioactive waste created by the nuclear process . . .

She also argues that "while currently the creation of nuclear electricity produces only one-third the amount of CO_2 emitted from a similar-sized, conventional gas generator, this is a transitory statistic. Over several decades, as the concentration of available uranium ore declines, more fossil fuels will be required to extract the ore from less-concentrated ore veins." Caldicott acknowledges that reprocessing fuel could recover large amounts of radiation but argues this process is too dangerous to workers and the environment (Caldicott 2007).

Assuming such problems could be overcome, just how many new nuclear plants would be needed to bring about a significant reduction in global warming? According to Richard Lester, head of the Nuclear Science and Engineering Department at MIT:

The IAEA recently made a projection of what would be needed to prevent an increase of more than 2° C in the global average surface temperature by the end of the century. There are many uncertainties in these kinds of calculations, but the bottom line of the IAEA projection is that the global nuclear fleet would have to be meeting about a quarter of the world's electricity needs by 2050 to achieve this goal. This would require an installed capacity of about 1200 GW, compared with 370 GW today. In other words, we would have to build more than twice as many nuclear plants in the next forty years as we built in the last 40. And since much of the current nuclear fleet will likely have been retired before the year 2050, the requirement for new build would be larger still. (Lester 2013)

Lester acknowledges that this would be a formidable task buts suggests the power of innovation so amply demonstrated in other technologies as well as specific developments such as small modular reactors could make this goal achievable.

Renewables as an Alternative

Another major argument against nuclear power as a solution to climate change is that renewable forms of energy (water, solar, and wind) can increasingly displace fossil fuels in the energy mix, avoiding the need for expensive and dangerous nuclear plants. Generally nuclear advocates have argued that renewables have their place but cannot provide the large steady "baseline" power supply that a modern civilization needs:

If the fundamental opportunity of these renewables is their abundance and relatively widespread occurrence, the fundamental challenge, especially for electricity supply, is applying them to meet demand given their variable and diffuse nature. This means either that there must be reliable duplicate sources of electricity beyond the normal system reserve, or some means of electricity storage. (World Nuclear Association 2012)

Actually, the particular tradeoffs vary with the source. Photovoltaic solar cells can be placed where power is needed, such as a house or public building. Wind turbines, on the other hand, require a good steady supply of wind, and the power they generate generally must be stored and transmitted to users.

The traditional power distribution system relies on a concentrated energy source (such as fossil fuels or nuclear) generating power from a centralized location and distributing it over the grid. These traditional system can provide a steady "base load"—the amount of power needed to meet anticipated demands throughout the day and night. A coal plant can adjust to demand by burning more fuel, while a nuclear reactor's output can also be adjusted.

With sun or wind, the maximum power generated is limited by what nature provides. Such power systems are thus often viewed as supplemental to traditional sources. However, Benjamin Sovacool, an energy policy and development expert, suggests that this situation is changing as renewable sources are linked into larger, more robust networks:

> Previously intermittent sources such as wind and solar can also displace nuclear resources. No less than nine recent studies have concluded that the variability or intermittency of wind and solar resources becomes easier to manage the more they are deployed and interconnected, and not the other way around, as some utilities suggest. (Sovacool 2011, 220)

Further, even as the capacity of solar power systems is increasing, a significant tax on carbon emissions would make it more cost competitive with fossil fuels, as admitted by the World Nuclear Association:

> With solar input being both diffuse and interrupted by night and by cloud cover, solar electric generation has a low capacity factor, typically less than 15%, though this

is partly addressed by heat storage using molten salt. Power costs are two to three times that of conventional sources, which puts it within reach of being economically viable where carbon emissions from fossil fuels are priced. (World Nuclear Association 2012)

The case of Germany may be illustrative of those who are choosing to move away from nuclear energy, even in the face of climate change. Europe's leading economy has been working to phase out nuclear power in favor of renewables. In a study of considerations for nuclear phase-outs, Alexander Glaser studies how Germany's phase-out decision developed, starting with the country's vigorous antinuclear movement of the 1970s. Glaser acknowledges that in countries with a significant nuclear energy share, "replacing nuclear power with other energy sources will require the phase-in of renewable energy sources at a faster rate than previously anticipated, and the accelerated transition is likely to come with a price tag." Nevertheless, Glaser shows how a strong and durable consensus for nuclear phase-out was crafted and that even growing concerns about global climate change were not affecting that decision, only important questions of timing. (Glaser 2012)

Germany has heavily invested in renewable power, increasing its electricity share from 6 percent in 2000 to 20 percent in 2011. These heavy subsidies are part of what the Germans call an Energiewende ("energy turn," or revolution). The goal is to reduce greenhouse emissions 40 percent from 1990 levels by 2020, and 80 percent by 2050.

However, Germany had also been relying on nuclear to provide 20 percent of its green energy. Now that the nation is phasing out nuclear, it is planning extensive wind farms along the North Sea coast. To keep the grid flowing reliably, the country is trying to develop massive power storage facilities to even out the fluctuating yield of wind and solar.

Graham Weale, chief economist for RWE, Germany's largest utility, says that "this Energiewende is being watched very closely. If it works in Germany, it will be a template for other countries. If it doesn't, it will be very damaging to the German economy and that of Europe" (quoted in Talbot 2012).

Finally, what does the public think of nuclear power as a climate solution? A 2009 study by Accenture Corporation found that when asked "What actions should be considered to reduce your country's reliance on fossil-fueled power generation (i.e., coal, oil, or gas generated power)?" only 9 percent favored increasing nuclear power, 34 percent favored increases in both nuclear and renewable energy, and 57 percent called for increasing renewable energy only. According to a survey of studies from Great Britain, "the researchers found that people were concerned about climate change, but radioactive waste trumped climate change in dread. There was also great mistrust of the competence of the nuclear-power establishment and the government to manage nuclear power safely" (Ramana 2011).

Part of the Solution?

If nuclear power is not a "magic bullet" for dealing with the climate crisis, might it still be a necessary part of a broader effort? After all, such measures as carbon taxes or massive subsidies for renewables also face major economic and political challenges. Suppose no one measure can reverse the buildup of greenhouse gases that threaten worldwide and potentially devastating climate changes?

In 2004 Stephen Pacala and Robert Socolow published an article introducing a new way to tackle the growth of carbon emissions and resulting climate effects. The goal is to "stabilize" the CO_2 level at about 500 parts per million (give or take 50), or roughly double its level before widespread industrialization. This would require holding annual emissions to 2004 levels of 7 billion tons of carbon per year, rather than allowing the current trend that would result in a doubling of carbon output.

The key insight is that relying on just one or two technologies or policies to stabilize carbon production results in a level of investment and commitment that would seem unrealistic to most observers. Instead, Pacala and Socolow divide the task of preventing 7 billion tons a year of additional carbon production into seven portions, or "wedges." These are called wedges both because they are like pieces of a pie and because like a wedge, each one starts out small and increases until it accounts for a billion tons of carbon. However, to actually reduce the carbon that has accumulated up to this time, emissions must be reduced below current levels to achieve stabilization.

Pacala and Socolow suggest that wedges can be achieved "from energy efficiency, from the decarbonization of the supply of electricity and fuels (by means of fuel shifting, carbon capture and storage, nuclear energy, and renewable energy), and from biological storage in forests and agricultural soils." They suggest that the greatest potential to achieve wedges comes from improved efficiency and conservation, including such measures as less use of cars and more fuel efficient vehicles, more efficient buildings (heating, lighting, etc.), improved power plant efficiency, and substituting natural gas for coal (something that may already be happening thanks to declining natural gas prices).

One of the other options listed in this category is nuclear fission power. To achieve a wedge (reduction of 1 billion tons of carbon), Pacala and Socolow estimate that 700 GW of the base load of power plants, currently burning coal would have to be replaced with new nuclear plants. This would be about twice the current total capacity of nuclear plants. They admit that "substantial expansion in nuclear power requires restoration of public confidence in safety and waste disposal, and international security agreements governing uranium enrichment and plutonium recycling." (Pacala and Socolow 2004)

Conclusion

If anything can be learned from this survey of issues relating to nuclear power, it is that it is not only the technology itself that is complex. Besides understanding (in broad outline at least) how nuclear plants work and the implications of various design decisions, evaluating nuclear power also requires a grasp of energy economics, the analysis of complex systems, risk analysis, workplace culture, popular opinion, geopolitics, and climate change.

In the end, a series of difficult decisions must be made by policy makers and ultimately, voters and consumers. What should be the priority of such goals as energy security, accommodation of future demand, and stabilization of the climate? Can nuclear power be made safe and economical enough to provide a significant amount of carbon-free energy? What is the appropriate "energy mix" between fossil fuels, renewable energy sources, and nuclear power? Or should nuclear power be phased out? If it is, will there be the political will to replace it with sufficient nonfossil power? Where do conservation and efforts to improve energy efficiency in vehicles, homes, and businesses fit in? It seems certain that whether it grows, shrinks, or stays the same, nuclear power will be an important issue for generations to come.

References

Abbot, Derek. 2012. "Limits to Growth: Can Nuclear Power Supply the World's Needs?" *Bulletin of the Atomic Scientists.*

Aldy, Joseph E. 2013. "Eliminating Fossil Fuel Subsidies." Brookings Institution. http://www.brookings.edu/~/media/research/files/papers/2013/02/thp%20budget%20papers/thp_15waysfedbudget_prop5.pdf

Baker, P. J., and D. G. Hoel. 2007. "Meta-Analysis of Standardized Incidence and Mortality Rates of Childhood Leukemia in Proximity to Nuclear Facilities." *European Journal of Cancer Care* 16 (4): 355–363. http://onlinelibrary.wiley.com/doi/10.1111/j.1365-2354.2007.00679.x/abstract

Biden, Joseph. 2010. "The Path to Nuclear Security: Implementing the President's Prague Agenda." Remarks at the National Defenses University, February 18, 2010. http://www.whitehouse.gov/the-press-office/remarksvice -president-biden-national-defense-university.

Black, Richard. 2011. "Nuclear Power 'Gets Little Public Support Worldwide.' " BBC News (November 24). http:// www.bbc.co.uk/news/science-environment-15864806

Bullis, Kevin. 2009. "Q & A: Stephen Chu." *MIT Technology Review* (May 14). http://www.technologyreview.com/news/ 413475/q-a-steven-chu/

Caldicott, Helen. 2007. *Nuclear Power Is Not the Answer*. New York: New Press.

Clinton, Hillary. 2009. "Secretary of State Hillary Clinton Remarks at the United States Institute of Peace," October 21, 2009. http://www.state.gov/secretary/rm/ 2009a/10/130806.htm

Cooper, Mark. 2012. "Nuclear Safety and Affordable Reactors: Can We Have Both?" *Bulletin of the Atomic Scientists* 68 (July/August).

Davenport, Coral. 2013. "Nuclear Waste in the Age of Climate Change." *National Journal* (March 2).

Department of Energy. 2007. "Department of Energy Five Year Plan, FY 2008–FY 2012." http://energy.gov/sites/prod/ files/em/EMFYPFinal4-6.pdf

Deutch, John M., et al. 2009. "Update of the MIT 2003 Future of Nuclear Power." http://web.mit.edu/ nuclearpower/pdf/nuclearpower-update2009.pdf

Dininny, Shannon. 2013. "6 Tanks at Hanford Nuclear Site in Wash. Leaking." AP (February 22). http://bigstory.ap.org/ article/gov-6-underground-hanford-nuclear-tanks-leaking-0

Du, Yanbo, and John E. Parson. 2009. "Update on the Cost of Nuclear Power." MIT Center for Energy and Environmental

Policy Research. http://web.mit.edu/ceepr/www/
publications/workingpapers/2009-004.pdf

Ebinger, Charles K., and John P. Banks. 2011. *Business and
Nonproliferation: Industry's Role in Safeguarding a Nuclear
Renaissance*. Washington, DC: Brookings Institution Press.

Elliott, A. (ed.). 2011. *COMARE 14th Report Further
Consideration of the Incidence of Childhood Leukemia around
Nuclear Plants in Great Britain*. http://www.comare.org.uk/
press_releases/documents/COMARE14report.pdf

Energy Information Administration. 2012. "Levelized
Cost of New Generation Resources in the Annual Energy
Outlook, 2011." http://www.eia.gov/forecasts/aeo/
electricity_generation.cfm

Energy Information Administration. 2013. "AEO2013 Early
Release Overview." http://www.eia.gov/forecasts/aeo/er/
early_elecgen.cfm

ExternE. 2006. "Final Reports of the Project." http://www
.externe.info/externe_2006/exterpols.html

Ferguson, Charles D. 2011a. *Nuclear Energy: What Everyone
Needs to Know*. New York: Oxford University Press.

Ferguson, Charles D. 2011b. "Think Again: Nuclear Power;
Japan Melted Down, But That Doesn't Mean the End of the
Atomic Age." *Foreign Policy* (November).

Folbre, Nancy. 2011. "Renewing Support for Renewables."
New York Times (March 28). http://economix.blogs
.nytimes.com/2011/03/28/renewing-support-for
-renewables/

Glaser, Alexander. 2012. "From Brokdorf to Fukushima: The
Long Journey to Nuclear Phase-Out." *Bulletin of the Atomic
Scientists* 68 (6).

Gore, Al. 2006. Speech on climate policy at the NYU Law
School (September 18).

Hvistendahl, Mara. 2007. "Coal Ash Is More Radioactive Than Nuclear Waste." *Scientific American* (December 13). http://www.scientificamerican.com/article.cfm?id=coal-ash-is-more-radioactive-than-nuclear-waste

Indiviglio, Daniel. 2011. "Why Are New U.S. Nuclear Reactor Projects Fizzling?" *Atlantic* (February 11). http://www.theatlantic.com/business/archive/2011/02/why-are-new-us-nuclear-reactor-projects-fizzling/70591/

Institute of Nuclear Power Operators. 2004. "Principles for a Strong Nuclear Safety Culture." http://www.efcog.org/wg/ism_pmi/docs/Safety_Culture/Dec07/INPO%20PrinciplesForStrongNuclearSafetyCulture.pdf

International Atomic Energy Agency. 1970. Treaty on the Non-Proliferation of Nuclear Weapons. http://www.iaea.org/Publications/Documents/Infcircs/Others/infcirc140.pdf

International Atomic Energy Agency. 2008. "Nuclear Security Culture Implementing Guide: IAEA Nuclear Security Series No. 7, 2008." http://wwwpub.iaea.org/MTCD/publications/PDF/Pub1347_web.pdf

International Atomic Energy Agency. 2012.

"Design Basis Threat (DBT)." http://www-ns.iaea.org/security/dbt.asp

Kaku, Michio. 2008. *Physics of the Impossible*. New York: Anchor.

Lester, Richard. 2013. "The Energy Question." *Nuclear Engineering International* (January).

Lovelock, James. 2005. "Our Nuclear Lifeline." *Readers Digest* (March).

Maeda, Risa. 2011. "Japanese Nuclear Plant Survives Tsunami, Offers Clues" (October 19). http://www.reuters.com/article/2011/10/20/us-japan-nuclear-tsunami-idUSTRE79J0B420111020

Metz, Bert (ed.). 2011. *Climate Change 2001: Mitigation; Contribution of Working Group III to the third assessment report of the Intergovernmental Panel on Climate Change.*, New York: Cambridge University Press, 2001.

Montopoli, Brian. 2011. "Poll: Support for New Nuclear Plants Drops." CBS News (March 22). http://www.cbsnews .com/8301-503544_162-20046020-503544.html

Norton, Boyd. 1982. "The Early Years." In *Nuclear Power: Both Sides*, edited by Michio Kaku and Jennifer Trainer. New York: Norton.

Nuclear Regulatory Commission. 2012. *State-of-the-Art Reactor Consequence Analyses (SOARCA) Draft Report, January 2012.* http://pbadupws.nrc.gov/docs/ML1202/ML120250406.pdf

Nuclear Energy Institute, 2014. "Nuclear Energy Statistics." http://www.nei.org/Knowledge-Center/Nuclear-Statistics/

Obama, Barack 2009. "President Barack Obama Delivers Remarks at Suntory Hall, Tokyo, Japan." *CQ Transcripts*, November 14, 2009.

Obama, Barack. 2012. "Remarks by President Obama at Opening Plenary Session of the Nuclear Security Summit." http://www.whitehouse.gov/the-press-office/2012/03/26/ remarks-president-obama-opening-plenary-session-nuclear -security-summit

Pacala, S., and R. Socolow. 2004. "Stabilization Wedges: Solving the Climate Problem for the Next 50 Years with Current Technologies." *Science* (August 13): 968.

Performance, Safety, and Health Associates. 2003. "Safety Culture Evaluation of the Davis-Besse Nuclear Power Station" (April 14). http://ocw.mit.edu/courses/nuclear -engineering/22-091-nuclear-reactor-safety-spring-2008/ readings/MIT22_091S08_read04.pdf

Ramana, M. V. 2011. "Nuclear Power and the Public." *Bulletin of the Atomic Scientists* 67 (4).

Rickwood, Peter, and Peter Kaiser. 2013. "Fast Reactors Provide Sustainable Nuclear Power for 'Thousands of Years.'" International Atomic Energy Agency. http://www.iaea.org/newscenter/news/2013/fastreactors.html

Rincon, Paul. 2013. "Nuclear fusion milestone passed at U.S. lab." BBC News October 7, 2013. http://www.bbc.co.uk/news/science-environment-24429621

Rosner, Robert, Rebecca Lordan, and Stephen Goldberg. 2011. *Bulletin of the Atomic Scientists* 67 (4).

Schneider, Mycle and Antony Froggatt. 2013. *World Nuclear Industry Status Report, 2013.* http://www.worldnuclear report.org/IMG/pdf/20130716msc-worldnuclear report2013-lr-v4.pdf

Schneider, Mycle, Antony Froggatt, and Steve Thomas. *Nuclear Power in a Post-Fukushima World.* Worldwatch Institute, 2011. http://www.worldwatch.org/system/files/pdf/WorldNuclearIndustryStatusReport2011_%20FINAL.pdf

Smith, Rebecca, 2010. "Small Reactors Generate Big Hopes." *Wall Street Journal* (February 18).

Sokolski, Henry (ed.). 2010. "Nuclear Power's Global Expansion: Weighing Its Costs and Risks." Strategic Studies Institute. http://www.strategicstudiesinstitute.army.mil/pubs/download.cfm?q=1041

Sovacool. 2011. *Contesting the Future of Nuclear Power.* Hackensack, NJ: World Scientific Publishing.

Szondy, David. 2012. "Small Modular Nuclear Reactors: The Future of Energy." *Gizmag* (February 16). www.gizmag.com

Talbot, David. 2012. "The Great German Energy Experiment." *Technology Review* (June 18).

Union of Concerned Scientists. 2007. "Nuclear Power and Global Warming." http://www.ucsusa.org/assets/documents/nuclear_power/npp.pdf

Union of Concerned Scientists. 2012. "Monitoring Nuclear Power: Flood Risk" (includes map of affected sites). http://www.ucsusa.org/nuclear_power/reactor-map/embedded-flash-map.html

University of Chicago. 2004. "The Economic Future of Nuclear Power" (August). http://web.archive.org/web/20070415012109/http://www.anl.gov/Special_Reports/NuclEconSumAug04.pdf

United Nations, 2005. "International Convention for the Suppression of Acts of Nuclear Terrorism: Article 2." http://treaties.un.org/doc/db/Terrorism/english-18-15.pdf

Wald, Matthew L. 2010. "Nuclear 'Renaissance' Is Short on Largess." Green blog. *New York Times* (December 7). http://green.blogs.nytimes.com/2010/12/07/nuclear-renaissance-is-short-on-largess/

Wald, Matthew L. 2013. "Ex-Regulator Says Reactors Are Flawed." *New York Times* (April 8). http://www.nytimes.com/2013/04/09/us/ex-regulator-says-nuclear-reactors-in-united-states-are-flawed.html

Wolfson, Richard. 1993. *Nuclear Choices: A Citizen's Guide to Nuclear Technology.* Cambridge, MA: MIT Press.

World Nuclear Association. 2012. "Renewable Energy and Electricity." http://www.world-nuclear.org/info/Energy-and-Environment/Renewable-Energy-and-Electricity/#.UUSmaHHvWkB

Wright, Kristen (ed.). 2012. "Nuclear Roundtable: Does U.S. Have Nuclear Future?" *Electric Light & Power* (May/June).

Yucca Mountain Project. 1999. "Total System Performance Assessment Peer Review Panel: Final Report" (February 11). http://energy.gov/sites/prod/files/edg/media/FinalRept-TotalSystemPerformanceAssessment-PRP.pdf

The use of nuclear power has been the subject of controversy for decades and has reheated in recent years as the public seeks alternatives to fossil fuels in response to the changing climate. This chapter provides a forum for six essayists with distinct viewpoints, defending some of the most common views on these topics. The collection opens with a spirited defense of the safety of nuclear energy relative to other energy sources and industries by noted scientist Dr. Brett McCollum, who is joined in his support of continued nuclear power development by Bruno Comby, founder and president of the association of Environmentalists for Nuclear Energy. Taking a more cautionary tone are Dr. Lloyd Dumas, who focuses on the inherent dangers of mixing powerful technology with the potential for human error, and Dr. Mark Troy Burnett, who points to many of the health and environmental risks of nuclear power generation. Hilary Olivia Faxon takes a constructive approach, drawing on her research with the Yale School of Management and arguing for the continued value of nuclear in our energy mix while pointing to important changes that should be made in the wake of the Fukushima disaster. Finally, rounding out the forum, Dr. Bruce E. Johansen offers an essential perspective

Demonstrations like this 2002 rally showed widespread opposition by environmentalists and local residents to the building of a permanent high level nuclear waste facility in Yucca Mountain, Nevada. (AP Photo/Dennis Cook)

to the discussion by examining the impact of uranium mining on the Navajo community.

Nuclear Energy Is a Safe Industry

Brett McCollum

Every time you step out of your home you are exposed to numerous hazards, each with an associated quantifiable risk. For every 10 miles you ride a bicycle you face a risk of fatality of 1 in 1 million (Langland et al. 2002). That same risk is connected to drinking 30 cans of diet soda or being a 60-year-old man for 20 minutes. Has that risk stopped people from riding bikes, drinking diet soda, or being 60 years old? Risk is a part of life. Societies aim to mitigate risk through laws that protect members from their own poor choices and those of others. We have laws proscribing acceptable driving behavior. We have laws governing the production and preparation of food. We have laws describing the delivery of medical treatment. With the objective to support the well-being and happiness of its members, a free and open society must ask what risks should be managed, which can be reduced, and if some risks should be avoided altogether. Like all industries, nuclear energy has significant associated risks. This essay explains why regulation frameworks and enforced regulatory authority make nuclear energy worth the risk.

The Worst Nuclear Energy Accidents in History

The nuclear industry is one of the most tightly regulated industries. Accidents in the atomic energy sector reach beyond one's national borders, both in terms of the possibility of released radiation as well as consequences for nuclear operations in other countries. However, relative to other energy sectors, nuclear energy has had few accidents per megawatt-hour of electrical production.

The three best-known nuclear energy accidents are: (1) Three Mile Island, (2) Chernobyl, and (3) Fukushima. All three of these events were the result of ineffective regulatory authority, violations, or nonadherence to prescribed safety protocol. Each has been followed by intense scrutiny of global nuclear activities and improvements in regulatory frameworks.

The Three Mile Island accident occurred in Pennsylvania on March 28, 1979. In this event the main water circulator pumps accidently went offline while all three of the emergency pumps were unavailable. These backup pumps had simultaneously been undergoing maintenance for two weeks, which was a considerable violation of the regulator's operating rules (Medvedev 1991; International Atomic Energy Agency 2013).

The Chernobyl disaster took place on April 26, 1986, near Pripyat in the Ukraine, formerly a part of the Soviet Union. In the midst of the Cold War, the Soviets were concerned about attacks on nuclear power plants from Western countries. A weakness was identified in the Chernobyl nuclear power plant cooling system, and an experiment was designed to test a possible solution without the need for an extensive retrofit. To conduct the test, the emergency system responsible for core cooling was disabled. This action was like overriding your home's fuse box, which helps protect you against electrocution. As the test progressed, all of the automatic (passive) safety systems were deactivated. Despite numerous alarms, control rods were removed from the reactor's core, including some of the "fail-safe" rods that were never meant to be removed and were meant serve to control the fission reactions as a last resort. The Organisation for Economic Co-Operation and Development's Nuclear Energy Agency (OECD-NEA) describes the Chernobyl disaster as an accident that "was the product of a lack of 'safety culture'" (OECD-NEA 2013). Additionally, the Chernobyl reactor lacked a containment dome, a design not used elsewhere because of safety issues.

On the International Nuclear and Radiological Event Scale, besides Chernobyl, the only other Level 7 event (major accident) occurred on March 11, 2011, at the nuclear power plant in Fukushima Prefecture, Japan. Reactors 1–3 at the Fukushima Daiichi power plant were online when the island country was rocked by a massive earthquake. Following protocol, these reactors were automatically shut down. Unfortunately, the quake had downed power lines to the facility. Without either internal or external power generation, the backup diesel generators were used to operate the water cooling system. While there was plenty of fuel to run the backup generators until outside assistance could arrive, a massive 13-meter (43 feet) tsunami struck the facility less than an hour after the earthquake, destroying the generators and leaving the plant without power to run the pumps (Lipscy et al. 2013). It had previously been identified that the existing 10-meter (33 foot) seawall was insufficient protection, but Japanese nuclear regulators had not been successful at requiring TEPCO (the utility company managing the power plant) to improve the barrier (Nishikawa 2013). Investigations have revealed a systemic problem within the Japanese nuclear regulator: large utility operators in Japan were experiencing accidents at a rate 42 percent higher per plant than smaller operators, possibly from poor enforcement of safety regulations (Lipscy et al. 2013).

Comparisons with Other Energy Sectors and Industries

All sources of energy have risks and inherent drawbacks. If nuclear energy is not used, what will fill the void? Coal, oil, and natural gas—as carbon-based energy sources—pump vast amounts of greenhouse gases (GHGs) into the atmosphere. Hydroelectricity is largely tapped out in North America, preventing its use for further production. Wind and solar energy require massive tracts of land to generate comparable levels of electricity, leading to poor efficiency and environmental footprints, thus reducing their impact for large-scale utility (Ausebel 2007).

While the aforementioned nuclear accidents have been alarming, industrial accidents in less regulated sectors are common and routinely claim more lives. On April 5, 2010, an explosion at the Upper Big Branch coal mine in West Virginia claimed the lives of 29 out of 31 miners. Fertilizer, an important energy source for food production, fuelled a massive explosion at the West Fertilizer Company in West, Texas, on April 17, 2013. Fifteen people were killed and 160 more were injured. With continued demand for petroleum products but insufficient pipelines to transport crude oil, railway has been used as a delivery mechanism. On July 6, 2013, a train transporting crude oil derailed in Lac-Mégantic, Quebec. Roughly half of the downtown was destroyed in the explosion. Forty-two people were confirmed dead, and five more are missing and presumed dead. One of the worst industrial accidents in history was the 1984 gas leak at a pesticide production facility in Bhopal, India. Over 3,700 people were killed from exposure to methyl isocyanate gas, while nearly 600,000 more experienced long-term suffering from gas-related illnesses. Even a lack of electricity has significant consequences in our modern society, as demonstrated by the Great Ice Storm of 1998. Thirty-five deaths were linked to the extreme weather.

Each of these events caused more deaths than Three Mile Island or Fukushima, yet our society continues to use coal and oil, fertilizer and pesticides. Populations continue to exist in regions that experience extreme weather. Scientists warn that extreme weather events will continue with greater frequency based on GHG production rates from coal, oil, and natural gas, and yet the human population consumes them at increasing rates. Nuclear energy has among the lowest greenhouse gas emissions in the energy sector, lower than wind and hydro and rivaling that of solar energy production.

The Benefits of Nuclear Energy

Nuclear reactors provide important benefits to society. First and foremost is the inexpensive low-GHG electricity. This is particularly noteworthy for nations with growing populations, such as India and China. New small modular reactors open up possibilities of providing nuclear energy to remote populations, such as those in Canada's Arctic, where the cost of electricity is more than 10 times that in major cities (Government of the Northwest Territories 2011). To put this in context, an average home in Calgary uses 7,200 kilowatt-hours of electricity each year, which does not include heating costs (City of Calgary 2013). This represents an annual cost of $575 for electricity. In Canada's Arctic that number would rise to around $5,750 per year, just for electricity. Heating expenses to survive the −28 °C (−18 °F) winter days using carbon-based fuels is separate from electricity costs. Low-cost nuclear energy could improve the financial freedom of these northern communities.

An often overlooked benefit of nuclear reactors is the production of medical isotopes. Some radioisotopes obtained from nuclear reactors, such as technetium-99m, are used for diagnostic imaging of the brain, heart, lungs, liver, thyroid, gallbladder, kidneys, and skeletal and circulatory systems. The vast majority of isotopes for medical diagnosis and treatment are produced at five reactors around the globe in Canada, Belgium, South Africa, the Netherlands, and France. The supply of isotopes for the United States and Canada comes almost exclusively from the National Research Universal Reactor (NRU) at Chalk River, Ontario, which represents over 1 million procedures each year. In 2007 the Canadian Nuclear Safety Commission ordered the shutdown of the reactor until the emergency power supplies for the reactor's cooling pumps were upgraded. The resulting isotope shortage caused by the unplanned shutdown prompted a unanimous vote in the Canadian House of Commons to restart the reactor immediately. Parliamentarians spoke about how the risk from a lack of medical isotopes outweighed the immediate need for a third emergency power supply for reactor pumps. The NRU is scheduled for decommissioning in 2016. Canada announced in 2012 their intention to instead produce technetium-99m from cyclotrons. The change in production techniques will make it no longer possible to ship medical isotopes to the United States. It is unclear at this time what plans the United States has in place to deal with the shortage expected in 2016. Meanwhile, Australia has expanded its nuclear capabilities for medical isotope production, ensuring a reliable 50-year supply for that continent.

Conclusions

The use of nuclear energy has associated risks. Good regulation, strong enforcement authority, and reliable operators will ensure that nuclear energy is a safe industry. When faced with a difficult choice, the Canadian Parliament voted unanimously to restart a nuclear reactor, demonstrating that the benefits of nuclear energy, as a source of inexpensive low-GHG electricity and medical isotopes, outweigh the risks of going without. Nuclear energy is clearly worth the risk.

References

Acton, J. M., and M. Hibbs. *Why Fukushima was Preventable.* Washington, DC: Carnegie Endowment for International Peace, 2012.

Ausebel, J. H. "Renewable and Nuclear Heresies." *International Journal of Nuclear Governance, Economy and Ecology* 1, no. 3 (2007): 229–243.

City of Calgary. "Energy Use in Your Home." http://www.calgary.ca/UEP/ESM/Pages/Reducing-Calgarys-ecological-footprint/Home-energy-savings/What-you-can-do/Energy-use-in-your-home.aspx, accessed October 1, 2013.

Government of the Northwest Territories. *Northwest Territories Energy Report.* 2011. Yellowknife, NT: Government of the Northwest Territories.

International Atomic Energy Agency. *Frequently Asked Chernobyl Questions.* 2013. http://www.iaea.org/newscenter/features/chernobyl-15/cherno-faq.shtml

Langland, O. E., R. P. Langlais, and J. Preece. *Principles of Dental Imaging.* Philadelphia: Lippincott Williams & Wilkins, 2002.

Lipscy, P., K. E. Kushida, and T. Incerti. "The Fukushima Disaster and Japan's Nuclear Plant Vulnerability in Comparative Perspective." *Environmental Science & Technology* 47 (2013): 6082–6088.

Medvedev, G. *The Truth about Chernobyl.* Translated by E. Rossiter. New York: I. B. Tauris, 1991.

Nishikawa, J. "TEPCO Found Wanting after Knowing Full Well the Tsunami Danger." *Asahi Shimbun*, June 3, 2013.

Organization for Economic Cooperation and Development -Nuclear Energy Agency (OECD-NEA). "Chernobyl: Assessment of Radiological and Health Impact; 2002 Update of Chernobyl, Ten Years On." https://www.oecd -nea.org/rp/chernobyl/c01.html, accessed October 1, 2013.

Brett McCollum *is associate professor in the department of chemistry at Mount Royal University in Calgary, Alberta. He holds a PhD from Simon Fraser University, where he studied chemical reaction mechanisms using the radioactive positive muon at the TRIUMF cyclotron (Canada's National Laboratory for Particle and Nuclear Physics). An important product of his research, the McCollum mechanism, demonstrated the first observation of a secondary muoniated radical in over 40 years of muon spin spectroscopy. McCollum has taught across the chemical spectrum, including courses on general chemistry, organic chemistry, inorganic chemistry, physical chemistry, and nuclear chemistry. He is responsible for the creation of three new courses and leading the development of course learning objectives in two others. His course*

"The Science and Politics of Nuclear Energy" is co-taught by Dr. Duane Bratt, chair of the Department of Policy Studies.

Safety, Human Fallibility, and Nuclear Power

Lloyd J. Dumas

Human fallibility is an unavoidable part of any human enterprise. We all know from our daily experiences that people make mistakes—even well-trained, careful people. And as the news repeatedly reminds us, there is always a subset of the population that is predisposed to malevolent behavior, be it criminal or terrorist in nature. Since fallible people inevitably interact with every technology we develop and deploy, we must be exceedingly careful to stay away from technologies that are capable of creating major disasters if enough goes wrong, by accident or intention. Unfortunately, nuclear power, which once fueled our hopes for a cheap and abundant source of energy, has turned out to be just such a technology.

Nuclear power presents us with three classes of serious safety problems: (1) the possibility of horrendous accidents, (2) vulnerability to terrorist acts, and (3) dangerous long-lived nuclear wastes.

The three worst nuclear power accidents that have occurred to date were at Three Mile Island (United States, 1979), Chernobyl (Soviet Union, 1986), and Fukushima (Japan, 2011). All of them involved major elements of human error and technical breakdown, along with the failure of backup systems intended to protect against error and breakdown. At Three Mile Island, there was a partial meltdown of the reactor core. Yet the presence of a containment structure ultimately prevented most of the radioactive material that would otherwise have been released from contaminating the surrounding land and threatening the lives of many thousands of people. At Chernobyl, operator error was compounded by design error—the reactor design was not only dangerously unstable, but there was also no containment—and enormous numbers of people were ultimately exposed as radioactive material drifted from Ukraine over Europe and beyond. Although events at Fukushima were precipitated by an earthquake and the tsunami it produced, human error played a central role. The people who designed and built the nuclear power stations there designed them to withstand earthquakes and other natural disasters. But they seriously underestimated the severity of the earthquake and tsunami those plants ultimately had to face. They weren't careless, and they weren't sloppy. They just made a mistake. But that's the point—people make mistakes. In this case, because it involved nuclear power, it was a very expensive mistake that radioactively contaminated a large area, seriously disrupted the lives of millions of people, and cost many billions of dollars.

Because it is a powerful impact multiplier, terrorists and other malcontents have long had nuclear power on their minds. U.S. Nuclear Regulatory Commission (NRC) data on "safeguards events" involving nuclear materials, power plants, and other facilities show almost 2,000 threatening events from 1976 to 2000, more than 40 percent of which were bomb related (Dumas 2010). There have been a number of serious incidents since then, including an attack by armed intruders who breached the control room at the Pelindaba nuclear plant (South Africa, November 2007). But it may be that the most serious attack terrorists have attempted came on September 11, 2001.

The only airliner hijacked on that day that did not reach its target was United Airlines flight 93. The plane flew from the East Coast heading west and slightly south. After the hijacking, it looped around and headed east again, ultimately crashing in Pennsylvania after the passengers and crew fought with the hijackers. Before it crashed, it was headed toward and only about 120 miles (193 kilometers), about 15 minutes flying time from the Three Mile Island nuclear plant. The NRC has repeatedly stated that the containments of American nuclear power reactors are designed to withstand the impact of a small airliner flying 150 miles per hour (240 km. per hour)—not a Boeing 767 coming in at 500-plus miles per hour (800 or more km. per hour). If that fourth hijacked plane had reached and crashed into the containment on the reactor building at Three Mile Island, we might well have had an American Chernobyl on our hands.

Nuclear waste is often corrosive and physically hot, as well as radioactively "hot." That makes it difficult to store safely isolated from the biosphere, especially when it remains dangerous for an extremely long time. A single half-life of plutonium, one of the elements in reactor waste and a critical component of nuclear weapons, is 24,000 years—much longer than all of recorded human history. About 1,000 metric tons of plutonium is contained in stored spent fuel rods from nuclear power plants, roughly four times as much as has been used in making all of the world's nuclear weapons.

When we built the first nuclear plants, we knew we didn't know how to neutralize or safely store the dangerous wastes they produced. But we were confident that well before it became a big problem, we would have developed effective waste treatment technologies. Now, some 60 years later, we have made a little progress. But we still do not have a safe, cost-effective way of dealing with nuclear waste. Yet even if no new nuclear power reactors were built and those currently in operation in the United States were retired at the end of their licensed operating life, we would still have 84,000 metric tons of spent commercial nuclear fuel to deal with.

If there were no safer alternative to nuclear power as a source of electricity, we might feel compelled to put up with its dangers and risks. But there are a whole set of environmentally sensible energy efficiency and renewable energy technologies available—some already on-line, some waiting just around the corner—that do not have the problems or disaster potential of nuclear power. For example, the worst possible accident or terrorist attack that could occur at a wind power facility might do a few million dollars worth of damage and cost one or two dozen lives. That pales beside the tens of billions of dollars and the tens of thousands of human casualties that we know similar disasters at a nuclear power plant can cause. And wind power doesn't generate any dangerous wastes. Nuclear power is simply too unforgiving a technology for a world filled with fallible human beings.

Reference

Dumas. Lloyd J. *The Technology Trap: Where Human Error and Malevolence Meet Powerful Technologies.* Santa Barbara, CA: Praeger, 2010.

Lloyd J. Dumas *is a professor of political economy, economics, and public policy at the University of Texas–Dallas. He is the author of* The Technology Trap: Where Human Error and Malevolence Meet Powerful Technologies (Praeger 2010).

Building Back Better: Lessons from Fukushima

Hilary Oliva Faxon

On March 11, 2011, a massive earthquake off the coast of Japan caused a series of powerful tsunamis that smashed into the Fukushima nuclear power plant, disabling its cooling system and resulting in subsequent nuclear meltdown and widespread radioactive contamination. Approximately 200,000 people were evacuated from the area. Over the months of complicated cleanup that followed, both Japan and Germany, countries with recent plans for major scale-ups of their nuclear capacity, loudly disavowed the energy source. Cutbacks in these two nations drove a global nuclear generation decrease of 4.3 percent, the largest on record (BP, 2012). It seemed long-controversial nuclear power had been dealt a decisive blow.

Or maybe not. A special report by the World Energy Council on the first anniversary of the Fukushima accident declared that the incident had hardly impacted energy policies outside of Japan, Germany, Italy, and Switzerland. Strong support for nuclear in energy-hungry China and India—as well as in Vietnam, the Middle East, Central Europe, the United Kingdom, and France—implies that nuclear is far from off the table (World Energy Council 2012). Despite previous disasters like those at Three Mile Island and Chernobyl, the public has largely grown to accept nuclear, and even including these accidents, the industry safety record in terms of per capita deaths and injuries is far better than those of fossil fuels such as coal and oil (see Table 2, Nuclear Energy Agency 2010). In the United States and in a world stricken by climate change, nuclear power continues to be a strong source of clean-air energy.

While the "nuclear renaissance" may be indefinitely stalled, nuclear energy, along with enhanced renewable technology and energy efficiency efforts, will continue to be a part of our clean energy efforts and global energy mix for the next several decades. A number of pre-Fukushima concerns still exist: reactor safety, waste management, nuclear weapon proliferation, social fear and acceptance. But Fukushima also brought to light new challenges, and potential solutions, for creating safer nuclear power. Some of these lessons are outlined in the following text.

1. Technical Improvements

The extreme magnitude of the natural disasters impacting the Fukushima nuclear facility exceeded what the system had been designed to handle; key technical improvements will recognize and plan for the possibility of significant disruption. A report from the Massachusetts Institute of Technology's Nuclear Science and Engineering faculty identifies key areas of design improvement. These cover issues of emergency power and response, containment, hydrogen management, spent fuel pools, and plant siting and layout considerations. Many of the recommendations are both simple and powerful. For example, house back-up generators at higher elevation and store transportable generators nearby (Fukushima's diesel generators were located in the basement of a coastal building and so quickly flooded and failed) (Buongiorno et al. 2011). The U.S. Nuclear Regulatory Commission issued three tiers of recommendations for improving facility safety that address similar technical issues.

Beyond specific enhancements, a critical concept is designing for redundancies and reinforcements that allow the system to remain functioning, or at least contained, after an initial failure.

2. Disaster Preparedness and Planning

Disaster management starts with prevention. Proper siting and facility layout can significantly reduce risk and damage. Fukushima powerfully demonstrated that nuclear facilities, like all other infrastructure, are embedded in their physical environment and are subject to local shocks and strains. Inadequate sea walls enabled the plant's flooding, as they were designed to prevent damage only from waves one-third the size of those seen in the event. Siting of other nuclear facilities near or on seismic fault lines demonstrates a prevailing ignorance to physical surroundings and vulnerabilities that must be changed in the wake of this accident.

A thoughtful, public disaster plan that clearly delineates protocol and chains of command can encourage preparedness among staff and provide guidance at the time of crisis. These plans must extend beyond the initial disaster to containment, cleanup, and follow-up. An evacuation plan for the surrounding area and agreed-upon radiation standards could have significantly reduced the chaos of the weeks following the Fukushima accident, in which successively larger groups of people were forcibly evacuated with little notice. By accepting the possibility of unexpected disruptions, prevention and planning reduce their likelihood and enable a more effective response.

3. Good Governance

"What must be admitted—very painfully—is that this was a disaster 'Made in Japan.' Its fundamental causes are to be found in the ingrained conventions of Japanese culture: our reflexive obedience; our reluctance to question authority; our devotion to 'sticking with the program'; our groupism; and our insularity" (National Diet of Japan 2012, 9).

While these characteristics may be especially strong in Japanese culture, lack of whistleblowers is a major factor in escalation of all sorts of crises, from the 2008 global financial crisis to the 2010 BP oil spill in the Gulf of Mexico. In Fukushima, the individual heroism of one plant manager who disregarded central orders limited the extent of the disaster. The power of the individual in warning of weaknesses and responding to disturbance cannot be underestimated, and promoting this personal critique and investment can strengthen existing systems and crisis response.

But relying on personal sacrifice and initiative is not enough—regulatory and governance systems must be designed to encourage scrutiny and safety. Energy producers and distributors often have massive market and political power. In Japan, the Tokyo Electric Power Company's tight relationship with the government allowed for breaches in safety standards and poor management at the time of crisis, including suppression of a document outlining worst-case scenarios for nuclear failure that might have helped mitigate impacts. Independent, transparent, and strong regulation of power companies is essential to maintaining safety and preventing corruption and negligence.

While the Fukushima nuclear plant failures did not cause any direct fatalities, the radioactive release and subsequent mass evacuation displayed the frightening vulnerabilities of this and other nuclear systems. Yet freezing nuclear in Japan has had major economic, climate, and energy security consequences as domestically generated energy is replaced with imported fossil fuel. Scientific models find that for national economics, the environment, and human safety, "no nuclear" fairs far worse than government plans that include nuclear in the energy mix (Hong et al. 2013). Contemporary politicians realize this as well: after shutting all 50 of its nuclear reactors following the disaster, Japan restarted two in July 2012, but after maintenance shutdowns, by the end of 2013 no nuclear power plants were operating in Japan. In a warming world with increasing demand for energy services, nuclear continues to play a small but essential part of our energy mix. With key lessons drawn from Fukushima, including technological improvements for reinforced systems, a new mindset of disaster preparedness, and requirements for transparent and accountable management, we can ensure a safer source of clean power.

References

BP. "Nuclear Energy." *BP Statistical Review of World Energy, June 2012.* http://www.bp.com/sectiongenericarticle800.do?categoryId=9037152&contentId=7068614, accessed 5/23/13.

Buongiorno, J., R. Ballinger, M. Driscoll, B. Forget, C. Forsberg, M. Golay, M. Kazimi, N. Todreas, and J. Yanch. *Technical Lessons Learned from the Fukushima-Daichii Accident and Possible Corrective Actions for the Nuclear Industry: An Initial Evaluation,* May 2011. http://web.mit.edu/nse/pdf/news/2011/Fukushima_Lessons_Learned_MIT-NSP-025.pdf, accessed 5/20/13.

Hong, Sanghyun, Corey J. A. Bradshaw, and Barry W. Brook. "Evaluating Options for the Future Energy Mix of Japan after the Fukushima Nuclear Crisis." *Energy Policy* 56 (May 2013) 418–424.

National Diet of Japan. *The Official Report of the Fukushima Nuclear Accident Independent Investigation Commission, Executive Summary*, 2012, http://www.nirs.org/fukushima/naiic_report.pdf, accessed 5/25/13.

Nuclear Energy Agency, Organisation for Economic Co-operation and Development, *Comparing Nuclear Accident Risks with Those from Other Energy Sources*, 2010. http://www.oecd-nea.org/ndd/reports/2010/nea6862-comparing-risks.pdf, accessed 5/29/13.

U.S. Nuclear Regulator Commission. "Implementing Lessons Learned from Fukushima." http://www.nrc.gov/reactors/operating/ops-experience/japan-info.html, accessed 5/23/13.

World Energy Council. *World Energy Perspective: Nuclear Energy One Year after Fukushima*, 2012. http://www.worldenergy.org/documents/world_energy_ perspective __nuclear_energy_one_year_after_fukushima_world _energy_council_march_2012_1.pdf, accessed 5/26/13.

Hilary Oliva Faxon *holds a Master of Environmental Management from Yale School of Forestry and Environmental Studies, where her research focused on energy and environmental governance in Asia and the developing world. She has published and presented on the impacts of new democracy on the environmental policy of the Himalayan kingdom of Bhutan, with a focus on impacts of the expanding hydropower sector. She is currently creating a case study on the Fukushima accident for a Yale School of Management course on managing global catastrophes. Past projects include economic analysis of shifting urban residential energy use in China, monitoring and evaluation on local and regional water and sanitation projects in rural Haiti, and a comparative analysis of political, economic, and safety of* *nuclear power in France, the United States, and the developing world.*

Nuclear Power: A Faustian Bargain

Mark Troy Burnett

Nuclear energy is a dangerous Faustian bargain; an overhyped cornucopia that while living up to its peril has never lived up to its promise. Despite the hype promulgated by the industry and even by progressive, environmental stalwarts such as Al Gore, the putative renaissance of nuclear energy has faulty premises: economically, technically, environmentally, and socio-politically.

The resurgence of attention on nuclear energy is directly related to concern with global warming and climate change. Though, quite frankly, nuclear energy needs climate change more than climate change needs nuclear energy. The nuclear industry and its advocates have—in my opinion, quite immorally—taken advantage of the world's concern with global warming and greenhouse gases to promote nuclear energy as a more environmentally friendly "carbon"-free option. This twenty-first-century public relations turn is misleading, and to some degree, just plain wrong. The building and eventual decommissioning of nuclear power stations is not carbon free; the mining, extraction, and transportation of fissile materials is not carbon free; the disposal and storage of radioactive waste is not carbon free. Would your average household want its nearby power station generating carbon dioxide, which isn't toxic and in truth is one of the least potent of the greenhouse gases, or plutonium, which, while not a greenhouse gas, is one of the most toxic substances on earth as well as the source for the most powerful, city-annihilating weapons? Furthermore, even if the risks were acceptable, the nuclear option is not cost effective or profitable. Indeed, the free market and its investment capitalists continually eschew nuclear energy in favor of renewable energy technologies (Lovins 2009).

The Problems with Nuclear: The Risks and Dangers Massively Outweigh the Benefits

To viscerally know the dangers of nuclear power one only has to reflect on Japan's Fukushima Daiichi disaster, which began in 2011. The continuing saga has dramatically unfolded on live television, revealing that even with precautions and advanced engineering safety mechanisms and systems, disasters can and do happen. While the initial accident was not human error or malfunctioning reactors but rather an earthquake and resultant tsunami, the fact remains that the Daiichi reactor exploded, releasing radioactive gases into the surrounding towns and environment. Not surprisingly, the nuclear industry has been denounced for this accident. Germany has shut down seven of its oldest facilities; Switzerland announced that it is reassessing its nuclear program and has suspended plans to replace existing reactors; and even China has stopped its plans for all nuclear plants and halted all plants in construction for the foreseeable future (Sovacool 2012).

Doubtless, the ecological and economic consequences of such accidents are devastating and enduring. The Ukrainian Chernobyl nuclear plant, which melted down in 1986 as a result of both technological and human failure, is presently encased in a massive cement sarcophagus that will have to be monitored and maintained for hundreds of years to avoid further radiation leaks. More than 350,000 people had to be forcibly resettled from the region and the ghostly, abandoned village of Prypiat serves as a stark reminder to the consequences of accidents.

Though the industry repeatedly claims that nuclear accidents are quite rare, Benjamin Sovacool has reported that globally, there have been 99 serious accidents at nuclear facilities, including the most dramatic examples of Windscale (1957), Idaho Falls, SL-1 (1961), Three Mile Island (1979), Tokaimura

(1999), and the aforementioned Chernobyl (1986). Astoundingly, this is almost two per year since nuclear energy became a serious energy provider in the 1950s, and 56 of the 99 have occurred in the United States. Effectively, the Fukushima disaster puts the number at an even 100.

According to Sovacool:

> If 10 million people were exposed to radiation from a nuclear meltdown, about 100,000 would die from acute radiation sickness in six weeks. About 50,000 would experience acute breathlessness and 240,000 would develop acute hypothyroidism. About 350,000 men would be temporarily sterile, 100,000 women would stop menstruating, 100,000 children would be born mentally disabled, and there would be thousands of spontaneous abortions and more than 300,000 cancers to develop later. In the U.S. presently, 80 million people live within 40 miles of a nuclear reactor. (Sovacool 2012)

The tragedy of a nuclear accident is the slow, persistent way it disrupts and destroys whole communities. Long after the immediate drama has passed, residents in and around the exposure zone continue to suffer. According to long-time industry journalist Stephanie Cooke:

> You have people in Japan right now that are facing either not returning to their homes forever, or if they do return to their homes, living in a contaminated area ... And knowing that whatever food they eat, it might be contaminated and always living with this sort of shadow of fear over them that they will die early because of cancer ... It doesn't just kill now, it kills later, and it could kill centuries later ... I'm not a great fan of coal-burning. I don't think any of these great big massive plants that spew pollution into the air are good. But I don't think it's really

helpful to make these comparisons just in terms of number of deaths. (Cooke 2011)

Nuclear power, according to Amory Lovins of the Rocky Mountain Institute, "is the only energy source where mishap or malice can destroy so much value or kill many faraway people; the only one whose materials, technologies, and skills can help make and hide nuclear weapons; the only proposed climate solution that substitutes proliferation, major accidents, and radioactive-waste dangers" (Lovins 2011). Even "normally" functioning reactors are associated with greater risk of unexplained human deaths and cancer. More than 100 radioactive chemicals (including strontium-90, iodine-131, and cesium-137) are produced by nuclear reactors. Tasteless, odorless, and invisible, strontium-90 has been linked to bone cancer, leukemia, and breast cancer and has been found in the teeth of babies living near nuclear power stations. The aforementioned plutonium is a by-product of fission and the preferred fuel for nuclear weapons. Its half-life is 25,000 years, and it is so dangerous that one pound spread evenly could give everyone on earth cancer. In 2009 the German Childhood Cancer Registry at the University of Mainz completed an intensive study on the relationship between proximity to nuclear power stations and incidences of childhood cancers and leukemia. The study revealed a "strong" relationship, such that children living within five kilometers (three miles) are twice as likely to develop leukemia (Nussbaum 2009).

A major environmental issue is water: its supply and usage and as a potential source of political conflicts. The nuclear industry has a vast appetite for water, with serious consequences for human and environmental health. All stages of the nuclear fuel cycle—facility construction, operation, and waste storage—consume, deplete, and contaminate water. Even the process of uranium mining itself is very wasteful and hazardous. A typical fission reactor requires 25 tons of uranium, which when dug out of the earth produces 500,00 tons of waste rock;

120,000 tons of poisonous mill tailings; and 1,400 cubic meters of liquid waste—all toxic for hundreds of thousands of years. Plant operation requires massive amounts of water to cool reactor cores and fuel rods. Not surprisingly, in comparison to other energy-generating facilities, a nuclear plant uses more water. A typical wind farm, for instance, requires 200 gallons per megawatt-hour; solar energy, 300 gallons per megawatt-hour; hydroelectric energy, 3,700 gallons per megawatt-hour—all this compared to nuclear energy's whopping 47,000 gallons per megawatt-hour. In addition, nuclear power plants dramatically alter the temperatures of nearby lakes, rivers, and streams as well as leak radioactive tritium and other toxic substances into nearby groundwater sources. Sovacool notes that as of February 2010, twenty-seven of the 104 reactors operating in the United States had been shown to be leaking tritium into watersheds (Sovacool 2011).

Another unresolved issue with nuclear energy involves the decommissioning of outdated facilities and the storage of the radioactive waste, which, it bears repeating, will be toxic and radioactive for tens of thousands to millions of years. The short-term strategy of the industry and its supporting governments has been to store the bulk of the waste near the existing facilities or deposit it at aboveground facilities throughout their territories, usually in the least inhabited or poorest regions. The long-term proposals involve deep geologic placement. These proposals have encountered many technical problems and have yet to be implemented. Primarily, the issue is finding a suitable, stable, economically and politically viable site. As any geologist will tell you, the ground beneath our feet is hardly stable; it moves, cracks, shifts, and has a sizable flow of groundwater that can and will be contaminated by the nuclear waste unless the perfect site is chosen. In the United States, the proposed facility at Yucca Mountain, Nevada, was recently rejected by the Obama administration for the primary reason that it could not guarantee containment beyond 10,000 years. The EPA's minimum requirement is 1 million years. Even if approved, the site would not have the capacity to store the existing radio-active waste, let alone what would be produced if we continue along the nuclear path. With centralized storage sites, there is the issue of moving the country's waste from its existing, dispersed locations. One can picture hundreds of trains, trucks, and planes loaded with highly toxic materials crisscrossing the country—everyone hoping an accident won't occur.

Conclusion

The arguments against assuming the risks of nuclear energy are quite clear. Furthermore, there are clear alternatives. Along with increasing efficiencies and appropriately matching the scale of energy usage, there needs to continue to be a serious move toward renewable, low-impact sources (solar, wind, co-generation, and geothermal) to the point where those technologies can achieve economies of scale with the present fossil fuel–based system and heavily subsidized nuclear option. This is what is known as the soft energy path and has proven to have a much lower environmental impact (Lovins 2011).

As most societies move toward a more sustainable energy future, the nuclear option continues to persist. I believe there are six socio-political conditions that best explain this situation: (1) a strong state involvement in guiding economic development, (2) centralization of national energy planning, (3) public relations campaigns linking technological progress to a national revitalization, (4) influence of technocratic ideology on policy decisions, (5) subordination of challenges to political authority, and (6) low levels of civic activism (Sovacool 2011). Global warming and climate change are also being used by the industry and its advocates to keep the option on the table. As an active global civil society, citizens and stakeholders must confront these conditions if we are finally going to be able to move beyond the nuclear option to a more sustainable energy plan.

Faust lost everything that was dear to him with his deal with the devil, and everyone is familiar with the chaos and destruction wreaked by Dr. Frankenstein's hubristic creation. In short, don't be fooled by the allure—nuclear energy is dangerous, ecologically destructive, and a red herring to distract us from better solutions.

References

Cooke, Stephanie. "Japan Says It Was Unprepared for Post-Quake Nuclear Disaster." *Los Angeles Times*, June 8, 2011.

Lovins, Amory. *Nuclear Nonsense*. Boulder, CO: Rocky Mountain Institute, 2009.

Lovins, Amory. *Soft Energy Paths for the 21st Century*. Boulder, CO: Rocky Mountain Institute, 2011.

Nussbaum, Rudy. "Childhood Leukemia and Cancers near German Nuclear Reactors: Significance, Context, and Ramifications of Recent Studies." *Journal of Occupational Environmental Health* 15, no. 3 (2009): 318–323.

Sovacool, Benjamin K. *Contesting the Future of Nuclear Power: A Critical Assessment of Atomic Energy*. Hackensack, NJ: World Scientific, 2011.

Mark Troy Burnett *is currently an associate professor of geography and the geography program coordinator at Mount Royal University in Calgary, Alberta. He holds a PhD in geography from the University of California–Los Angeles and a bachelor's degree from the University of California–Santa Barbara, in economics and environmental studies. His areas of academic interest are political, economic, environmental, and cultural geography.*

Why Nuclear Energy Is Clean and Environmental Opposition to Nuclear Energy Is a Major Mistake

Bruno Comby

I find it paradoxical that today many environmental groups are opposed to nuclear energy. Their announced concerns are for health, safety, and the protection of nature. In these respects nuclear energy is superior to the alternatives—burning fossil fuels (coal, oil, and gas), using solar photovoltaic cells and wind turbines for the production of electricity, and biomass (burning crop residues and growing crops to be burned).

Well-designed, well-constructed, well-operated, and well-maintained nuclear energy is clean, safe, reliable, durable, and competitive. Let's discuss each point.

Clean: Pollution and the Greenhouse Effect

About 30 billion tons of carbon dioxide (CO_2) are produced every year and are dumped into the atmosphere, producing the well-known greenhouse effect. Similarly, millions of tons of sulfur dioxide—which produce acid rain and nitrogen oxides—cause respiratory difficulties. Nuclear power produces little carbon dioxide and neither sulfur dioxide nor nitrogen oxides.

There are those who have fallen in love with the simplicity of solar cells and especially the pristine elegance of wind turbines, but they refuse to accept the observation that wind and solar are intermittent and quantitatively incapable of supplying the energy required by an industrial civilization. I do not mean to say that these renewable energies should be excluded. In specific cases they can play a large role, to some extent. My own house, which I built with my own hands in Houilles, in the close suburbs of Paris, is ecologically powered by solar energy and is energy positive (i.e., it produces more power each year than my family requires over the entire year). But this energy is not produced when it is required, only when the wind blows or the sun shines. Therefore, even the best ecohouse (such as mine) still needs to be connected to a reliable source of power that is available on demand, as large quantities of electricity cannot be stored and must be produced when needed. Wind and solar are useful and have important niche roles to play in remote locations, on our rooftops, and under special circumstances, but they can make only partial contributions to the energy demands of an industrial civilization.

Fields of solar cells, wind turbine farms, and growing biomass—all occupy large areas, and wind turbines disfigure the countryside and still can't fulfill our energy requirements. A nuclear power station is compact, typically occupying the area of a football stadium and its surrounding parking lots.

Safety

Nuclear power is safe, as proven by the record of half a century of commercial operation, with the accumulated experience of more than 10,000 reactor-years. There have been three serious accidents in the commercial exploitation of nuclear power: Three Mile Island 2 (1979, Pennsylvania), Chernobyl 4 (1986, Ukraine, Soviet Union), and Fukushima (March 2011, Japan).

Three Mile Island was the worst accident one can imagine: the core of the reactor melted, and much of it fell to the bottom of the reactor vessel. But the radioactive fission products were almost entirely confined in the reinforced concrete containment structure, the airtight silo-like building that houses the reactor itself. It was designed for that purpose. A small and innocuous quantity escaped; no one was seriously irradiated, and no one died.

Chernobyl was different. The reactors at Chernobyl had no containment structure. Reactor 4 exploded, and its graphite moderator caught fire and burned for several weeks. The smoke carried radioactive fission products high into the atmosphere, where they were swept hither and yon by the winds. Forty-two workers died, and over 200 more were severely irradiated but survived. People living nearby were hurriedly uprooted, evacuated, and resettled elsewhere. They lost their jobs and homes, and suffered psychological and social trauma in the dissolving Soviet Union. Their lives were disrupted and certainly shortened. Some 1,800 cases of thyroid cancer were found in the nearby population, and most were successfully treated. It is not clear whether all these cancers were caused by irradiation or whether they were spontaneously occurring thyroid cancers detected by the enhanced screening and detection methods applied in the affected region.

Fukushima was a natural catastrophe, a typical illustration of the geologist's aphorism that *civilization exists by grace of geological stability, subject to cancellation without notice*. The Fukushima event was caused by a magnitude 9.0 earthquake followed by a 13-meter (43 foot tsunami), far higher than had been seen in that area before. The nuclear reactors shut down as they were supposed to, but auxiliary power supplies failed to operate. Four men died in the power station—a crane operator who was up in his crane at the moment of the earthquake, two workers who drowned in a flooded compartment, and one who died of over-exertion. Some 28,000 Japanese died as a result of the tsunami, which flooded a large area of the Pacific coast of Japan.

In sum, less than 50 fatalities have occurred in the civilian nuclear power industry in half a century, considerably fewer than occur in any single year in the fossil fuel electrical power industry. Coal mine accidents are common and often cause tens or hundreds of fatalities, reported one day and forgotten the next. The same may be said for oil field accidents. Oil tankers go aground or break up, accidents occur in refineries, and oil and gas platforms have been lost with all hands. Accidents in high-pressure gas pipelines are not infrequent. The gas pipeline accident at Ghislenghien (Belgium) on July 30, 2004, killed at 24 and injured 150.

Reliable

The sun shines only half the time at best, and the wind blows only when it will. They are hardly suited to furnish the base load power needs 24/7 for our industrial economy and urban civilization. Only fossil fuels and nuclear can supply those demands.

Conservation

We are urged to conserve energy. Conservation is indeed highly commendable, even essential, especially for those advanced countries that now depend on massive imports of fossil fuels. But in the face of the growth of the world's population and its enhanced expectations, notably China and India (about 35 percent of the world's population), and in the face of finite fossil fuel resources, conservation can only delay the crisis by a few years or a few decades. (And then there are those who seek a "simpler" life, but that question is beyond the concern of this essay. As commendable as it may be, if people do so, this may sweep us back into the middle ages, a time when our fellow humans' main energy was their hands and legs, without machines to help with tedious tasks—for example, at that time women had to walk long distances to the river to clean their family's clothes by hand in cold water . . .)

Nuclear Waste Is a Political Problem

A gram of uranium yields about as much energy as a ton of coal or oil—thus the famous "factor of a million" effect. Nuclear wastes are accordingly about a million times smaller than fossil fuel wastes. Most fossil fuel waste goes up the smokestack and we don't see it, but it is not without effect, causing global warming, acid rain, smog, and other atmospheric pollution.

The volume of nuclear waste produced is very small, it is totally contained, and it can be stored safely. Its radioactive components decay spontaneously. In the United States, spent fuel is simply stored away. Elsewhere, it is reprocessed to separate the radioactive fission products and heavy elements (about 3 percent), which are vitrified for safe and permanent storage, and to recover the plutonium and uranium-238, which are made into new fuel elements and thus recycled to produce more energy. The La Hague reprocessing plant near Cherbourg (France) is the largest installation of this kind and reprocesses spent fuel for Japan, Germany, and other countries as well as France.

Proliferation

The only serious argument against nuclear power is the fear of nuclear weapons proliferation, that is, that some enriched uranium or separated plutonium may fall into mischievous hands and be fashioned into a bomb. No state now possessing nuclear weapons reached that condition via nuclear power, but we are now dealing with at least one inimical nonstate entity whose only access to nuclear weapons material would be theft. Eternal vigilance is indeed imposed upon us.

An intelligent combination of energy conservation and nuclear energy for base load electricity production and renewable energies for local low-intensity applications (with nuclear and hydraulic energy as a reliable backup) is the way of the future.

The environmental movement's opposition to civilian applications of nuclear energy will in the future be revealed as among the greatest mistakes of our times.

Nuclear power is necessary. It is needed. It is clean, safe, cheap, and available when and where it is needed—on demand. It is absolutely required to simply ensure the survival of our industrial civilization when oil and gas start running short a few decades from now. We have no choice. There is no other solution. In fact, we are very lucky to have this convenient and safe nuclear solution so that we may avoid the disappearance of our modern civilization.

Bruno Comby *is the founder and president of the association of Environmentalists For Nuclear Energy (http://www.ecolo.org). He is the author of 10 books on healthy living, which have been published in 15 languages. He is a graduate of the Ecole Polytechnique in Paris and holds a postgraduate qualification as* *nuclear physicist from ENSTA, the National University of Advanced Technology in Paris.*

Navajos and Uranium Mining: Death Digging Yellow Dirt

Bruce E. Johansen

The Navajo language has no word for "radioactivity." In the late 1940s when men in suits came to the reservation offering mining work, many Navajo men had no idea that the search for yellow dirt would kill them. The nuclear industry unleashed a fatal plague in Navajo country to fuel a growing power base and military arsenal. Today hundreds of abandoned mines (so small that workers often called them dogholes) lay silent. Uranium mining and milling is now illegal in Navajo Country, regarded as a crime against the people and against nature. This is as it should be. Visiting the mining sites, ghosts call out.

The mining of uranium first became a major political issue for the Navajo during the 1970s as many miners began to die of lung cancer. Activists in the Coalition for Navajo Liberation sought to have uranium mining and milling outlawed on the reservation as soon as its human toll became evident (50 miners had died by 1980), a position that eventually prevailed in the Navajo government.

The number of deaths increased steadily after 1980. Lung cancer had been all but unknown among the Navajo, but mining caused a plague of it. Lung cancer results from inhalation of radon gas, a by-product of uranium's decay into radium. Miners who worked for Kerr-McGee during the 1940s were exposed to between 100 and 1,000 times the dosage of radon now considered safe by the federal government. Harris Charley, who worked in the mines for 15 years, told a U.S. Senate hearing in 1979, "We were treated like dogs. There was no ventilation in the mines." Pearl Nakai, daughter of a deceased miner, said at the same hearing, "No one ever told us about the dangers of uranium" (Grinde and Johansen 1995, 214).

For several years after the first atomic bombs were tested in New Mexico, uranium was mined much like any other mineral. More than 99 percent of the mined product was waste and was cast aside as tailings near mine sites after the uranium had been extracted. One of the mesa-like waste piles grew to be a mile long and 70 feet high (1.6 km. by 21 meters high). On windy days, dust from the tailings blew into local communities, filling the air and settling into the water supplies. No health, safety, or pollution regulations existed. As Navajo miners continued to die, children who played in water that had flowed over or through abandoned mines and tailing piles came home with burning sores. Workers loaded the radioactive ore into wheelbarrows, emerging from the mines spitting black mucus from the dust and coughing so hard it gave many of them headaches. Relatives of the dead recalled how the miners had eaten their lunches in the mines, washing them down with radioactive water, never having been told that it was dangerous.

Uranium—"yellow dirt"—is part of the natural order of things in Navajo Country, and a ban on its mining and milling was regarded by traditional people as a cultural victory, a return to the original instructions of creation. The yellow dirt in mythology is the antithesis of the life-giving yellow powder of corn pollen.

> In one of the stories the Navajos tell about their origin, the *Dineh* (the people) emerged from the third world into the fourth and present world and were given a choice. They were told to choose between two yellow powders. One was yellow dust from the rocks, and the other was corn pollen. The Dineh chose corn pollen, and the gods nodded in assent. They also issued a warning. Having chosen the corn pollen, the Navajo were to leave the yellow dust in the ground. If it was ever removed, it would bring evil. (Eichstaedt 1994, 47)

In this cultural interpretation, the commercial use of the yellow dirt turned it into *Leetso*, a powerful monster that inflicted punishment upon the Navajo people in the form of disease and suffering. The only proper cultural response in Navajo tradition was to restore harmony with nature, which is a very important part of the Navajo way, *hozho nashaadoo*, "to walk in harmony" (Csordas 1999). Harmony could be restored only by banning the mining and milling of uranium. Some of the elders blamed themselves for allowing disruption of hozho nashaadoo (Brugge and Benally 2006, 106).

Another aspect of restoration was compensating victims. Bills to compensate the miners were introduced and discussed, and they died in the U.S. Congress for a dozen years. By 1990 the death toll among former miners had risen to 450 and continued to rise. Once compensation finally was approved, requirements were made so stringent that very few miners qualified. As years passed, more died.

Healing of the earth is another way to restore harmony. This is a monumental task. The nuclear mining legacy of 60 years still blows through the outlying districts of Shiprock, New Mexico, the largest Navajo town, on windy days. The hot, dry winds shave radioactive dust from the tops and sides of large tailings piles around the town. After almost three decades of appeals from he Navajo Nation, the U.S. Environmental Protection Agency in September 2011 approved plans to clean up the Northeast Church Rock Mine, near Gallup, New Mexico, the largest of the many abandoned uranium mines on or near the Navajo Nation.

Healing has barely begun. The mining companies and the U.S. government face a billion-dollar task, one that humanity and morality demand they complete.

References

Brugge, Doug, and Timothy Benally. *The Navajo People and Uranium Mining*. Albuquerque: University of New Mexico Press, 2006.

Csordas, T. J. "Ritual Healing and the Politics of Identity in Contemporary Navajo Society." *American Ethnologist* 26 (1999): 3–23.

Eichstaedt, Peter. *If You Poison Us: Uranium and American Indians.* Santa Fe, NM: Red Crane, 1994.

Grinde, Donald A., Jr., and Bruce E. Johansen. *Ecocide of Native America: Environmental Destruction of Indian Lands and Peoples.* Santa Fe, NM: Clear Light, 1995.

MacMillan, Leslie. "Uranium Mines Dot Navajo Land, Neglected and Still Perilous." *New York Times*, March 31, 2012.

Pasternak, Judy. *Yellow Dirt: An American Story of a Poisoned Lands and a People Betrayed.* New York: Free Press, 2010.

Bruce E. Johansen *is the Jacob J. Isaacson Professor in Communication and Native American Studies at the University of Nebraska at Omaha. He has authored or edited 38 books, the most recent of which is* The Encyclopedia of the American Indian Movement *(Greenwood, 2013). In addition to writing on Native American studies, Johansen also has written widely on environmental studies.*

A number of individuals and organizations have played a key role in the development of nuclear power and the debate over its risks, safety, and future. More information on most individuals and organizations can be found by searching online sources.

People

Bernard Baruch (1870–1965)

Wealthy as the result of successful stock speculation, Baruch played an important role in national politics starting with his chairmanship of the War Industries Board during World War I. In 1919 he served as an economic advisor on President Woodrow Wilson's team at the Paris Peace Conference. During the 1930s he served as an adviser on economic policies to President Franklin D. Roosevelt.

When the control of nuclear energy became a key issue in the aftermath of the development of the atomic bomb during World War II, Baruch played a significant role in the attempt to create a workable international framework that would both restrain the development of nuclear weapons by other nations

Equipped with a harness and other safety equipment, inspector Justin Heinly prepares to climb to the top of a reactor building at Three Mile Island. (Courtesy of NRC.gov)

and promote the peaceful development of nuclear power. What became known as the Baruch Plan for International Control of Atomic Energy suggested that the United Nations impose sanctions on countries that violated agreements on the use of atomic energy, while at the same time the United States would offer help in establishing nuclear power industries in cooperating nations.

Baruch's plan met with little acceptance. Most Americans wanted the United States to jealously guard its nuclear monopoly so that the Soviets could not develop nuclear weapons to counter those of the United States. The Soviets in turn opposed the Baruch Plan, claiming that its real purpose was to perpetuate the U.S. nuclear monopoly. By 1949 the Soviets had exploded their first atomic bomb and the nuclear arms race was under way. Baruch's autobiography, *Baruch*, was published in two volumes (Holt, 1957–1960).

Hans Albrecht Bethe (1906–2005)

Born in Strasbourg, France, and trained in Germany, physicist Hans Bethe went to England and then the United States following the rise of the Nazi Party to power during the 1930s (Bethe's mother was Jewish). During the war he worked on the development of both radar and the atomic bomb. (When some scientists worried that an atomic explosion might set off a runaway fusion reaction in the earth's atmosphere, Bethe's calculations helped dispel their fears.) His enduring legacy, however, is in advancing the understanding of the nucleus and nuclear reactions. In particular, he was able to determine the precise nature of the nuclear fusion reaction that powers stars like our sun, in which hydrogen is the fuel but carbon serves as a catalyst. He won the Nobel Prize in Physics in 1967 for this work. Remarkably, Bethe continued active scientific research and publication well into his nineties. Fellow physicist Freeman Dyson referred to Bethe as "the supreme problem solver of the 20th century."

In the social and political realm, Bethe worked tirelessly to promote nuclear disarmament, a ban on nuclear weapons tests, and initiatives for the peaceful use of atomic energy. He opposed Edward Teller's plans for the development of the hydrogen bomb, hoping that both the United States and the Soviet Union could agree to not build such superweapons. During the 1980s, when Teller promoted the Strategic Defense Initiative (also known as Star Wars), Bethe opposed the effort as technically impractical and politically destabilizing. In 1995 Bethe wrote an open letter asking all scientists to "cease and desist" from any involvement in the development or manufacture of nuclear weapons.

Niels Bohr (1885–1962)

After Bohr earned his PhD in physics in 1911 from the University of Copenhagen, Denmark, he became an assistant to James J. Thompson (discoverer of the electron) at Cambridge University, England, and then worked at Manchester University under Ernest Rutherford, where he helped him make his key discoveries about the nature of the atomic nucleus and the results of bombarding it.

Bohr's key contribution, however, was in developing the comprehensive theoretical understanding of the atom that would explain Rutherford's discoveries. In 1913 Bohr used the new quantum theory to explain the orbits of electrons around the nucleus; he is, in fact, considered to be the leading figure in early quantum physics. In 1922, Bohr won the Nobel Prize in Physics for his work.

During 1938 and 1939 Bohr increasingly turned his attention to the possibility of liberating vast amounts of energy through nuclear fission, which was emerging from the work of Enrico Fermi, Otto Hahn, and Lise Meitner. He came to the United States and warned scientists that Germany's Nazi Party might be developing atomic weapons. In 1943 the British air force helped Bohr escape from Nazi-occupied

Denmark. He joined the Manhattan Project at Los Alamos, New Mexico, where he provided theoretical help for the atom bomb effort. After the war, Bohr returned to Denmark and devoted himself to peaceful applications of atomic energy.

A good source for reading about Bohr's life and work is *Niels Bohr: His Life and Work as Seen by His Friends and Colleagues* (North-Holland, 1967).

Helen B. Caldicott (1938–)

Australian physician and antinuclear activist Helen Caldicott has brought intense energy and passion to the antinuclear movement. Her first involvement came in 1971 when she warned the public of dangerous fallout from French nuclear bomb tests in the South Pacific. The publicity helped lead to the election of an antinuclear Labor Party slate to the Australian parliament, but the movement was unsuccessful in banning exports of uranium from Australia (one of the world's major producers of the substance).

Caldicott moved to the United States in 1975 and continued her activism. Her book *Nuclear Madness: What You Can Do* (1978; revised ed., Norton, 1994) became influential in the growing public opposition to nuclear power and nuclear weapons. The same year she revived the organization Physicians for Social Responsibility and gave it a new antinuclear focus. She also founded the group Women's Action for Nuclear Disarmament in 1980. During the 1980s she played a major role in the revived opposition to nuclear weapons, and she also promoted general environmental concerns.

Her autobiography is entitled *A Desperate Passion* (Norton, 1996).

In 2001 Caldicott founded the Nuclear Policy Research Institute. In 2007 Caldicott was designated a Women's History Month honoree by the National Women's History Project. In 2009 she published a revised edition of *If You Love This Planet: A Plan to Save the Earth*. (W. W. Norton)

Jimmy Carter (1924–)

The thirty-ninth president of the United States, James Earl Carter was born in Plains, Georgia. Graduating from the Naval Academy at Annapolis in 1946, Carter later served aboard nuclear submarines and helped the outspoken Admiral Hyman Rickover in the navy's nuclear engineering program.

Carter's involvement with nuclear propulsion as a naval officer gave him more technical background in nuclear matters than most politicians. However, as president (1977–1981), Carter promoted an environmental agenda that minimized nuclear power, despite serious economic problems caused by the high price of oil. Under the Carter administration the commercial reprocessing and recycling of plutonium produced in power reactors was deferred, as was the development of breeder reactors. A greater emphasis was placed on the need to prevent the proliferation of sensitive technologies such as for uranium enrichment.

In foreign affairs he completed negotiation of the SALT II nuclear weapons limitation treaty, although its implementation was derailed following the Soviet invasion of Afghanistan. After he left office Carter became an internationally respected and effective negotiator seeking to resolve conflicts and promote human rights.

Bernard Cohen (1924–2012)

Physicist Bernard Leonard Cohen was a prolific writer on the health effects of radiation and wrote many scientific papers on risks associated with nuclear materials. A former president of the Health Physics Society, Cohen was an outspoken advocate of nuclear energy, asserting its fundamental safety and challenging the linear no-threshold model (LNT), which asserted that even tiny doses of radiation necessarily create incremental cancer risks. Cohen argued that the effort to prevent even tiny releases of radiation was costing billions of dollars that could be used to address more significant problems.

Further, Cohen argued in favor of a theory called radiation hormesis, which suggests that low-level radiation might actually be necessary and beneficial to organisms via stimulation of the cellular repair mechanism. However Cohen's views have remained a minority position rejected by organizations such as the U.S. Environmental Protection Agency and the World Health Organization.

Barry Commoner (1917–2012)

One of the most influential American environmentalists during the 1970s, Commoner, an accomplished zoologist, became a leading spokesperson against nuclear power. He concluded that the combination of inherent risks in the technology and influence of the industry over those who sought to regulate it makes nuclear power unacceptable.

Commoner's involvement in nuclear issues began in the late 1950s when he became part of a research team that showed nuclear testing was resulting in the accumulation of the radioactive isotope strontium-90 in children's teeth. Commoner became active in the movement to ban atmospheric nuclear testing. In 1970 he received the International Humanist Award from the International Humanist and Ethical Union.

Commoner wrote two influential books in the 1970s that helped form the modern environmentalist critique of energy policy and the movement for energy alternatives: *Alternative Technologies for Power Production* (Macmillan, 1975) and *Social Costs of Power Production* (Macmillan, 1975). His book *The Poverty of Power: Energy and the Economic Crisis* (Knopf, 1976) took a broad perspective on energy economics and energy choices for the future. He founded the Citizens' Party and ran for president of the United States on its ticket in 1980.

Commoner is remembered among ecologists for his four laws of ecology: everything is connected to everything else, everything must go somewhere, nature knows best, and there is no such thing as a free lunch.

Dwight D. Eisenhower (1890–1969)

Dwight D. Eisenhower, the chief commander of Allied forces in western Europe during World War II, emerged from the war as a popular hero and desirable political candidate. In 1952 he was elected U.S. president and served two terms.

Eisenhower's presidency coincided with the formative years of nuclear power in the United States, and he was a forceful advocate of its development. His "Atoms for Peace" speech was the keynote for the effort to turn the atomic energy that had first shown its face as a devastating weapon of war into a bountiful and reliable source of energy for America's booming economy. However, even as the foundations of the nuclear power industry were being laid, production of nuclear weapons greatly increased during Eisenhower's two terms in office, from about 1,000 to 20,000.

Eisenhower also sought to have nations join together to share and control nuclear technology. In 1957 the International Atomic Energy Agency was created to implement these ideas, and 62 nations signed the agency's charter. Proliferation of nuclear weapons was limited to only a few countries such as Britain and France. Domestically, Eisenhower's administration was marked by the establishment of the Atomic Energy Commission as both regulator and promoter of nuclear power in the United States, a dual role that—according to critics— would often prove problematic.

Many works have been written about Dwight Eisenhower. One good biography is *Eisenhower* by Stephen Ambrose, in two volumes (Simon and Schuster, 1983–1984).

Mohamed ElBaradei (1942–)

An Egyptian legal scholar, diplomat, and politician, Mohamed ElBaradei served as director general of the International Atomic Energy Agency (IAEA) from 1997 to 2009. Early in his first term ElBaradei strengthened the IAEA's ability to monitor nuclear facilities to safeguard against the diversion of nuclear materials to nonnuclear nations or terrorist groups. Following the September 11, 2001, terrorist attacks, ElBaradei and the IAEA stepped up the nuclear security program to assist member states in protecting their nuclear installations.

In ElBaradei's second term the agency's inspections in Iraq became embroiled in charges from the United States that Iraq was engaged in developing nuclear weapons. ElBaradei and Hans Blix led a team of inspectors to Iraq who concluded that charges that the regime had tried to obtain uranium from Niger were unfounded. ElBaradei became a strong critic of the U.S. invasion of Iraq in 2003, arguing that such actions only made control of nuclear materials more difficult. He also argued that a system that condemned some nations for pursuing nuclear weapons while others depended on them for security was ultimately unworkable.

By 2005 the United States was opposing ElBaradei's appointment to a third term, but most leading nations supported him, and he was ultimately elected. That same year ElBaradei and the IAEA were awarded the Nobel Peace Prize in recognition of their work. Following his departure from the IAEA, ELBaradei became involved in the 2011 Egyptian revolution and founded the new Egyptian Constitution Party in 2012.

Enrico Fermi (1901–1954)

Italian-born physicist Enrico Fermi, besides making important contributions to quantum physics, was a key researcher into the nature of radioactivity and atomic decay during the 1930s. He discovered element 93, neptunium, one of a series of transuranic elements that would play an important role in nuclear energy.

In 1938 Fermi went to Stockholm, Sweden, to receive the Nobel Prize for his work with radioactivity. Because of the dominance of Italy by Benito Mussolini's fascists, however, he emigrated to the United States rather than returning home. In 1942 he led the effort to build the world's first nuclear reactor beneath the football stands at the University of Chicago. He then worked on the atomic bomb at Los Alamos, New Mexico, and on the hydrogen bomb, although by 1949 he opposed further development of the fusion device.

Fermi later played a key role on the General Advisory Committee of the Atomic Energy Commission, and in 1954 he received its first special prize, which was given his name. The prize is for scientific and technical achievement in the development, use, and control of atomic energy. Emilio Segrè's biography *Enrico Fermi: Physicist* (University of Chicago Press, 1970) is a good introduction to Fermi's work and its significance.

John Gofman (1918–2007)

John Gofman's training as both a doctor of medicine and a PhD in nuclear chemistry equipped him well to play a pivotal role in establishing the modern science of health physics, which studies the effects of radiation on human health and develops ways to assess and reduce radiation exposure.

During World War II Gofman worked on the Manhattan Project and with Glenn Seaborg to develop two different methods for separating plutonium from irradiated fuel. This would lead to the development of the "plutonium factory" at the Hanford Nuclear Reservation in Washington State.

Gofman then returned to the academic world, where he had a long career as a professor of medical physics. He also served as director and administrator in the Lawrence Berkeley and Lawrence Livermore laboratories. Meanwhile, he continued his medical research. In addition to studying radiation effects, he did important work on lipoproteins and their connection to artery and heart disease.

Gofman's work with radiation led him to be increasingly critical of what he saw as lax standards and a lack of commitment from the nuclear establishment to protecting workers and the public from the dangers of radiation. He warned that the emerging medical consensus was that "there is no safe dose" of radiation and that it was important to avoid any unnecessary exposure. (Gofman supported the linear no threshold hypothesis that any amount of radiation exposure increased risk).

Gofman's research was attacked as "slipshod" by some critics, and his relationship with the AEC became strained. The agency cut off his research funding, but he found support for his work at Berkeley. Gofman testified in court on behalf of radiation victims in cases such as *Johnston vs. U.S.* and *Silkwood vs. Kerr-McGee*. A transcript of Gofman's oral history, "Human Radiation Studies: Remembering the Early Years," can be found at http://www.hss.energy.gov/healthsafety/ohre/roadmap/histories/0457/0457toc.html.

Henry M. Jackson (1912–1983)

Jackson, a Democratic U.S. senator from the state of Washington, had a long and distinguished career that spanned issues from national defense to labor and civil rights legislation. He served in the House of Representatives starting in 1940 and was elected to the Senate in 1953.

In 1955 Jackson proposed that President Eisenhower create a "bank" of nuclear materials not needed for weapons production that could be used for both military and civilian purposes. However, Eisenhower turned down the idea. Jackson naturally took a special interest in the huge Hanford, Washington, nuclear production facility. He gave his backing to the development of a dual-purpose nuclear reactor design that could produce plutonium for both weapons and electric power. Throughout his career Jackson remained a strong proponent of expanding nuclear power, including its use for generating electricity in remote areas such as Antarctica.

Abdul Qadeer Khan (1936–)

This prominent Pakistani nuclear scientist established Pakistan's uranium enrichment project and thus played a key role in the development of that nation's nuclear arsenal, although some experts consider his involvement to be overstated. Khan gained much of his expertise in gas centrifuges by working with the URENCO Group's uranium enrichment facility in the Netherlands.

In the mid-1970s, following Pakistan's defeat in the war with India and the latter's successful atomic bomb tests, Khan became involved with Pakistan's own bomb project. However, he became disenchanted with it because it continued to focus on plutonium rather than uranium as a fuel source. Khan's actual fame (or notoriety) would result from his developing contacts with China, North Korea, Iran, and Libya that involved sharing expertise in uranium enrichment. The degree to which these activities were known and approved by the Pakistani government remains in dispute.

Western governments began to respond to what they viewed as a serious nuclear proliferation issue. In 1983 a court in the Netherlands convicted Khan in absentia on espionage charges. Khan's attorney continued to contest the charges, arguing that the information involved was actually in the public domain. An appeals court eventually overturned the charges on a legal technicality. In 2004 Khan came under investigation by the Pakistani government and admitted to running a nuclear proliferation ring and transferring technology to North Korea and Libya. Rather than being arrested, Khan was "debriefed" on these activities by Pakistani authorities, and he was officially pardoned by President Musharraf later the same year.

The International Atomic Energy Agency and the U.S. government remained unsatisfied with this outcome and demanded that the IAEA be allowed to question Khan. However, these and efforts within Pakistan itself were quietly wrapped up with the fall of the Musharraf government.

Although criticized by many of his fellow scientists, Khan was able to continue to pursue his scientific career.

David E. Lilienthal (1899–1981)

An Illinois lawyer, Lilienthal became involved in the power industry when Governor Philip La Follette appointed him to the Wisconsin Public Service Commission. Later, President Roosevelt appointed him as one of three directors of the huge Tennessee Valley Authority (TVA). Lilienthal's insistence on separating the agency from politics earned him the enmity of some industrial interests.

In 1946 President Truman appointed Lilienthal as the first chairman of the Atomic Energy Commission, where he was largely responsible for defining the role and implementing the operations of the new agency within the guidelines set by the Atomic Energy Act. As he set to work to organize the agency and its key program for the development of nuclear power reactors, he tried to listen to public concerns and make the agency accountable. However, the growing Cold War fear of Soviet espionage and nuclear weapons led to an investigation of the AEC's security by Senator Bourke B. Hickenlooper and charges of mismanagement, which led to Lilienthal's resignation in 1949.

Lilienthal remained a thoughtful advocate of nuclear energy. His views can be found in his book *Atomic Energy: A New Start* (Harper and Row, 1980).

Lise Meitner (1878–1968)

Born in Vienna, Austria, Meitner obtained her doctorate in physics from Vienna University in 1906 and then went to Berlin, Germany, to a long-term collaboration with chemist Otto Hahn. In 1918 they discovered the element protactinium, and Meitner also did important research on radioactive decay. Despite the pervasive prejudice that erected formidable obstacles for women pursuing a career in science, Meitner persevered, becoming professor of physics at the University of Berlin in 1926. Adolf Hitler's rise to power, however, put increasing pressure on Meitner and other Jewish scientists in Germany, who were gradually stripped of their professional status.

Meitner's most important work came in the late 1930s, when she closely followed Hahn's work on the bombardment of heavy elements such as uranium. Studying these results, she came to the conclusion that uranium could fission, or split apart, into two approximately equal pieces, and she wrote the first paper describing the phenomenon.

Although she played no part in the development of the atomic bomb and little part in the subsequent development of nuclear power, Meitner's importance has been recognized in recent years. She did not receive a Nobel Prize (possibly because she was a woman and had worked in the background) but did receive many other honors. In 1997 chemical element 109 was named meitnerium in her honor.

A good biography is *Lise Meitner: A Life in Physics* by Ruth L. Sime (University of California Press, 1996).

Karl Z. Morgan (1907–1999)

Morgan, an American physicist, was influential in the development of the field of health physics, which studies the effects of radiation on humans and other organisms. Morgan had a long career at the very center of the American nuclear program and industry, including the wartime Manhattan Project and the Oak Ridge National Laboratory, where he became director of health physics.

After becoming a professor of nuclear engineering at the Georgia Institute of Technology (a position he left in 1982), Morgan gradually became a pronounced critic of the nuclear industry. During the 1980s Morgan testified on behalf of people who argued they had developed leukemia and other cancers due to exposure to the 1950s Nevada atmospheric nuclear weapons tests. Morgan also testified on behalf of Navajo who had worked as uranium miners and suffered health effects, as well as on behalf of the case of nuclear whistleblower Karen Silkwood against power producer Kerr-McGee.

Morgan's long and complex relationship with the nuclear age was chronicled in his autobiography *The Angry Genie: One Man's Walk through the Nuclear Age* (University of Oklahoma Press, 1999).

Mayumi Oda (1941–)

Japanese American silkscreen artist Mayumi Oda has brought together art, Zen Buddhism, peace activism, and strong feminine archetypes in her much-acclaimed work.

Oda grew up in Tokyo during World War II, experiencing the devastating firebombings and postwar hardships. During the 1960s—married to an American scientist and living in Cambridge, Massachusetts—Oda became involved with the growing women's movement. She then learned Japan was beginning an ambitious program of building breeder reactors to create fuel for a burgeoning nuclear power industry. For a time, she set her art aside in favor of antinuclear activism, helping found the group Plutonium-Free Future and coordinate the efforts of antinuclear organizations in Japan and North America.

Oda felt that the spread of nuclear power in Japan and throughout Asia might well fuel a nuclear arms race, and she argues that declining prices for uranium now make it economically unnecessary to produce plutonium fuel in breeder reactors. She also suggests that it is time for Asia and the world to turn to alternative forms of energy.

In recent years Oda has focused on her art, which has been widely shown and is influential.

J. Robert Oppenheimer (1904–1967)

Oppenheimer was born in New York City but received his PhD in physics from the University of Göttingen, Germany, in 1927. He performed important early research in quantum theory and the theory of antiparticles (particles, such as the positron, that are oppositely charged or oppositely spun from their normal counterparts).

From 1942 to 1945 Oppenheimer was director of the atomic bomb project at Los Alamos, New Mexico. Although he made key contributions to developing nuclear weapons, his experience of their devastating effects spurred him to oppose development of the more powerful hydrogen bomb and to advocate putting nuclear weapons under international control. From 1945 to 1952 Oppenheimer was chairman of the General Advisory Committee of the Atomic Energy Agency as well as a leading consultant to the U.S. delegation to the UN Atomic Energy Committee.

Oppenheimer's liberal views and suspected left-wing connections made him a target of the anticommunist forces of the McCarthy era. In 1953 he was suspended by the AEC as a security risk. He received much support from his academic colleagues, however, and in 1954 was unanimously reelected as the director of the Institute for Advanced Study at Princeton.

Readers interested in Oppenheimer's life and work have a variety of choices. One of the best is *J. Robert Oppenheimer: Shatterer of Worlds* by Peter Goodchild (Houghton Mifflin, 1981). This book was produced in conjunction with the BBC/WBGH television series Oppenheimer. A more recent and acclaimed biography is *American Prometheus: The Triumph and Tragedy of J. Robert Oppenheimer* by Kai Bird and Martin J. Sherman (Knopf, 2005).

Dixy Lee Ray (1914–1994)

After a long career as a marine biologist, Ray became the first female chair of the Atomic Energy Commission, serving from 1973 to 1975. She was a staunch advocate of nuclear power and had little patience for its critics. (She once pointed out that 300 people a year die just from choking on food, far more than have died in decades of nuclear power in the United States.) However, Ray was well aware of safety concerns, and she campaigned to eliminate defects in the construction and operation of nuclear power plants.

When the AEC was abolished in 1974 (to be replaced by the Nuclear Regulatory Commission), the nation was in the midst of the "energy crisis" that led many Americans to think seriously for the first time about energy alternatives and choices. Nuclear power, too, was about to suffer the shock of Three Mile Island and come under a firestorm of criticism. Leaving the federal government, Ray served as governor of Washington State from 1977 to 1981, having run as an independent. The idiosyncratic Ray bucked the tide of popular opinion in refusing to work to close the large nuclear production and waste facility at Hanford, Washington. She later became an independent consultant and wrote two books that clashed head-on with environmentalists: *Environmental Overkill* (HarperPerennial, 1994) and *Trashing the Planet* (Regnery Gateway, 1990).

Hyman G. Rickover (1900–1987)

Admiral Rickover played a key role in developing nuclear power as a source of propulsion for the U.S. Navy's submarines. He argued that with their ability to travel submerged indefinitely, such ships would be an essential tool for controlling the seas in future wars. His efforts culminated with the launching of the first nuclear submarine, *Nautilus*, in 1954. Rickover later became chief of the Naval Reactors Branch of the Atomic Energy Commission.

Rickover's outspoken, even badgering, style often involved him in controversy and acrimony, but his determination led to the triumph of the "nuclear navy" against entrenched old-guard opposition. Aircraft carriers and cruisers were also built with nuclear reactors. The success of the naval reactors proved to be an important spur to the development of civilian nuclear power, serving as proof of the concept that reliable reactors could be built and run for long periods without mishap.

In his later years Rickover became interested in educational reform and wrote several books on the subject. He received the Enrico Fermi Award in 1965 and the Medal of Freedom in 1980.

Glenn T. Seaborg (1912–1999)

Chemist Glenn T. Seaborg developed much of the understanding of the properties of transuranic elements (such as plutonium) that was necessary for the development of nuclear power. In 1951 he shared a Nobel Prize in Chemistry with Edwin McMillan for this work.

In 1961 President Kennedy appointed Seaborg as chairman of the Atomic Energy Commission. He served 10 years, the longest term of any AEC chairman. Seaborg believed that nuclear power was both technically and economically viable. He was an effective leader, combining the qualities of a scientist and a good administrator. He received the Enrico Fermi Award in 1959. In 1997 chemical element 106 was designated seaborgium in his honor.

Seaborg's book *The Nuclear Milestones* (W. H. Freeman, 1972) offers a good overview of atomic discoveries and the development of nuclear power.

Karen Silkwood (1946–1974)

Karen Silkwood became a cause célèbre of the antinuclear movement when she died in a car crash on November 13, 1974. She had been working as a chemical technician in the Kerr-McGee plutonium production plant at Crescent, Oklahoma. Prior to her death, she had been helping the Oil, Chemical and Atomic Workers' Union gather evidence about alleged unsafe conditions at the plant. She was exposed to plutonium under circumstances that remain mysterious. After she was killed as her car went off the road into a culvert, some investigators concluded that her car had been pushed from behind. When her body was examined, plutonium was found in her tissues. This suggested to nuclear opponents that she had been deliberately poisoned with the radioactive substance.

After Silkwood's death, her estate filed a lawsuit against Kerr-McGee, charging the company with maintaining unsafe working conditions. A jury awarded the estate $10.5 million. However, a federal appeals court reversed all but $5,000 of the award. As the estate prepared further appeals, the company settled out of court for $1.3 million. The Kerr-McGee fuel plant closed in 1975.

For more about the Silkwood mystery and case, see *The Killing of Karen Silkwood: The Story behind the Plutonium Case* by Richard L. Rashke (2nd ed. Cornell Paperbacks, 2000). Silkwood's story is also portrayed in an Academy Award–nominated movie, *Silkwood* (1983).

Lewis L. Strauss (1896–1974)

Lewis L. Strauss served in a variety of administrative capacities. During World War I he served in the Belgian Relief Commission under Herbert Hoover. During World War II he served as special assistant to Secretary of the Navy James Forrestal while attaining the rank of rear admiral.

In 1953 President Eisenhower appointed Strauss as chairman of the Atomic Energy Commission. He served simultaneously as Eisenhower's special assistant for atomic energy matters. His term on the AEC (1953–1958) was controversial. He engaged in a power struggle with the congressional Joint Committee on Atomic Energy over the role of the AEC and its relationship to the committee's oversight role. He also disagreed with J. Robert Oppenheimer, who did not share his strong support for developing the hydrogen bomb.

Strauss saw the role of the AEC as being more independent of Congress but also more limited in its scope. He believed that the federal government should play only a limited role in the Power Reactor Demonstration Program, which was designed to build demonstration reactors to show the commercial feasibility of nuclear power, and to help private industry start nuclear power programs. He believed that the government should not go beyond demonstration projects to the active subsidization of the nuclear industry.

Strauss's book *Men and Decisions* (Doubleday, 1962) recounts many of the issues he dealt with during his term on the AEC.

Leo Szilard (1898–1964)

Hungarian-born American physicist Leo Szilard was one of the central figures in bringing awareness of nuclear fission, the nuclear chain reaction, and its potential to the attention of the United States on the eve of World War II. After being first to describe the chain reaction, Szilard patented a design for a nuclear reactor. In 1939 Szilard was the actual author of the letter to Roosevelt under Einstein's signature that described the destructive potential of an atomic bomb and warned of the possibility that Nazi Germany may already have made substantial progress toward building such a weapon.

During the Manhattan Project, Szilard became disenchanted with the growing dominance of military over scientific considerations, and he argued unsuccessfully for conducting a demonstration of the bomb to convince Japan to surrender, rather than actually dropping it on a Japanese city. Following the war Szilard's misgivings about the use of nuclear energy led him to change fields entirely to molecular biology.

Grace Thorpe (1924–2008)

The daughter of famed Native American Olympic star Jim Thorpe, Grace Thorpe served in the Women's Army Corps in the Pacific in World War II. After graduation from the

University of Tennessee, Thorpe became a tribal district judge for the Fox and Sack Nation in Stroud, Oklahoma.

Thorpe became active in nuclear issues in 1992 when her tribe accepted a $100,000 grant from the federal government to study the establishment of nuclear waste dumps on tribal land. As the tribe's health commissioner, Thorpe organized a campaign to warn her people of the dangers of nuclear waste and reminded them of the many agreements with Native Americans that the government had broken in the past. The tribe then voted to reject the government's grant. Thorpe went on to organize declarations by 30 tribes from over 70 reservations who declared their lands to be "nuclear free zones."

Thorpe went on in 1993 to found the National Environmental Coalition of Native Americans (NECONA) and traveled around the country helping other tribes organize against nuclear waste.

Stewart Udall (1920–2010)

Stewart Lee Udall was part of a famous Arizona political family—his father, Levi S. Udall, was an Arizona Supreme Court justice, and his brother was former congressperson Morris Udall. In his youth Steward served as a Mormon missionary for two years and in the U.S. Air Force as a gunner during World War II. He began his long congressional career in 1954. He then served as secretary of the interior under Presidents Kennedy and Johnson.

Udall's legislative interests showed a major focus on protecting the environment, including preservation and expansion of parks and wilderness areas in his home state. During the 1970s he became a strong advocate of solar energy as a solution to the energy crisis. He also took up the issue of compensation for health effects suffered by uranium miners and their families, and in 1990 helped author the Radiation Exposure Compensation Act. He continued to address this and other aspects of the Cold War's nuclear legacy, as recounted in his book *The Myths of August: A Personal Exploration of Our Tragic Cold War Affair with the Atom* (Rutgers University Press, 1998).

Organizations

Alliance for Nuclear Accountability

Website: http://www.ananuclear.org/

This alliance of 35 organizations began in 1987 during a period of widespread concern about the continuing risks and environmental effects posed by nuclear weapons and associated activities. It focuses on trying to close unsafe weapons sites, production facilities, and waste storage facilities. More recently it has targeted efforts to reprocess fissionable material from obsolete nuclear warheads into fuel for power reactors and efforts to design a new generation of "reliable replacement warheads." Rather than just trying to clean up particular problems, the groups in this alliance target what they see as a corrupt system with inadequate regulation and oversight.

The alliance's website features news about recent events and a timeline of what it views as successful campaigns.

American Institute of Physics

Website: http://www.aip.org

This is an organization of scientific societies devoted to the advancement of physics, scientific cooperation, and public outreach and education. It organizes conferences and publishes periodicals and books (through the AIP press). Its Niels Bohr Library is an important source of archives for research into the development of nuclear fission, the atom bomb, and nuclear reactors. The AIP Emilio Segrè Visual Archives contain more than 25,000 photographs and other visual materials focusing on "the human face of science."

American Nuclear Society

Website: http://www.ans.org

Founded in the 1950s, the ANS is an international organization for advancing and coordinating the professions of nuclear science and engineering as well as promoting research and education. The ANS has 10,500 members drawn from 46 countries who work in diverse settings, including government agencies, academia, research laboratories, and private industry. The ANS has a wide array of publications, including journals such as *Fusion Science and Technology*, *Nuclear Science and Engineering*, and *Nuclear Technology*. Magazines include *Nuclear News* and *Radwaste Solutions*.

Argonne National Laboratory

Website: http://www.anl.gov

The Argonne National Laboratory near Chicago is run by the U.S. Department of Energy and has more than 1,250 scientists and engineers engaged in research involving energy, the environment, and national security.

The laboratory's Nuclear Engineering Division has been involved in the design and development of several research reactors. (Their predecessors built Chicago Pile 1, the world's first nuclear reactor.) It has also built reactors used in training nuclear technicians and operators. Today the laboratory has many research initiatives, including modeling, simulating, and developing new technologies.

Beyond Nuclear

Website: http://www.beyonduclear.org

This advocacy group was founded by the leading nuclear critic Dr. Helen Caldicott. Beyond Nuclear looks toward "a world free from nuclear power and nuclear weapons." It argues that nuclear power and weapons technologies are closely related and are inevitably unsustainable, environmentally pernicious, and undemocratic. Some of the topics covered on their website through a variety of media include reactor hazards and security risks, flaws in the planning and approval process, waste transportation and storage, and the impact of radiation on health. There is also extensive coverage of news involving every aspect of nuclear power, nuclear weapons, and the antinuclear movement.

Center for Nonproliferation Studies
Website: http://cns.miis.edu/

The James Martin Center for Nonproliferation Studies (CNS) at the Monterey Institute of International Studies provides research, news, and resources for professionals and the public to strengthen the effort to combat the spread of nuclear and other weapons of mass destruction. The organization conducts five research programs:

- Chemical and Biological Weapons Nonproliferation Program
- East Asia Nonproliferation Program
- International Organization and Nonproliferation Program
- Newly Independent States Nonproliferation Program
- Nonproliferation Education Program.

The organization publishes the *Nonproliferation Review* and a variety of papers.

Environmentalists for Nuclear Energy
Website: http://www.ecolo.org/base/baseen.htm

Founded by activist and prolific French author Bruno Comby in 1996, Environmentalists for Nuclear Energy (EFN) has about 10,000 members from 60 countries. The group argues that nuclear energy, properly managed, is safe and clean, and indeed the only practical energy source for reversing the ongoing degradation of the climate and environment. One prominent member is James Lovelock, the environmentalist and futurist who proposed the Gaia hypothesis of the biosphere as a self-regulating entity.

The EFN web site offers a variety of news stories, documents, and links to frequently asked questions about nuclear energy.

International Atomic Energy Agency

Website: http://www.iaea.org/

In the wake of President Dwight D. Eisenhower's 1953 "Atoms for Peace" address, the United States in 1954 proposed that an international "nuclear bank" be established to control the use of fissionable material and prevent the spread of nuclear weapons. The United States also proposed an international conference of scientists involved with the development of nuclear power.

While Soviet opposition doomed the nuclear bank, the scientific conference did take place in 1955. It was quickly followed by the organization of the International Atomic Energy Agency (IAEA), which began operations in 1957 and has its world headquarters in Vienna, Austria.

The IAEA is charged with fostering international cooperation on nuclear issues. IAEA groups in many countries are involved with the provision and supervision of nuclear fuel for power reactors and preventing the diversion of such materials for weapons purposes. The IAEA also monitors compliance with international treaties relating to nuclear materials and facilities, and reports regularly to the UN General Assembly and Security Council.

The IAEA frequently brings together nuclear scientists and engineers from around the world to address new developments in reactor technology and to develop or revise safety and security standards. The latter efforts were intensified following the 1986 Chernobyl nuclear disaster and again in response to the 2011 meltdowns in Fukushima, Japan. In 2005 the IAEA and its director, Mohammed ElBaradei of Egypt, were awarded the Nobel Peace Prize.

Nuclear Energy Agency

Website: http://www.oecd-nea.org/nea/

The Nuclear Energy Agency was organized under the Organization for Economic Co-operation and Development in 1958. (Originally the organization was called the European Nuclear Energy Agency, but in 1972 Japan joined, and "European" was dropped from the name.)

The NEA's stated mission is:

> To assist its member countries in maintaining and further developing, through international co-operation, the scientific, technological and legal bases required for a safe, environmentally friendly and economical use of nuclear energy for peaceful purposes. To provide authoritative assessments and to forge common understandings on key issues as input to government decisions on nuclear energy policy and to broader OECD policy analyses in areas such as energy and sustainable development.

Currently the NEA has 31 national members from Europe, North America, and the Asia-Pacific region, accounting for about 90 percent of the world's installed nuclear capacity. The NEA works closely both with its parent organization (the OECD) as well as the International Atomic Energy Agency (IAEA).

The organization has a strong emphasis on drawing on leading experts to develop useful and thoroughly researched technical reports.

Nuclear Energy Institute
Website: http://www.nei.org

The Nuclear Energy Institute was founded in 1994 via the merger of several industry organizations, including the U.S. Council for Energy Awareness, the American Nuclear Energy Council, and the nuclear division of the Edison Electric Institute.

Today the NEI is the principal public advocacy and public policy organization for the nuclear industry. The NEI website provides news, industry data and statistics, backgrounders and fact sheets, and policy papers.

In addition to power plants, the NEI represents other aspects of the nuclear industry, including nuclear medicine, agricultural and industrial use of radiation technologies, uranium mining and processing, and the transportation and storage of nuclear waste.

Nuclear Information and Resource Service
Website: http://www.nirs.org

NIRS was founded in 1978 as "the information and networking center for citizens and environmental organizations concerned about nuclear power, radioactive waste, radiation, and sustainable energy issues." In the ensuing decades the NIRS has become more activist, developing major organizing and public relations campaigns involving such issues as proposed new reactors, radioactive waste transportation, and what it views as the risks posed to the public by deregulation of low-level radioactive materials. The NIRS is affiliated with the international organization WISE (World Information Service on Energy).

Nuclear Threat Initiative

Website: www.nti.org

A nonprofit, independent organization dedicated to strengthening the Non-Proliferation Treaty by developing ways to reduce the spread and the risk of using nuclear and other weapons of mass destruction. The NTI works with organizations in many nations that face such threats and helps with efforts to secure nuclear stockpiles. The website provides detailed data on nuclear materials, trafficking, and other relevant statistics.

Oak Ridge National Laboratory

Website: http://www.ornl.gov/

The entire town of Oak Ridge, Tennessee, was built from scratch and in tight secrecy as part of the wartime Manhattan Project, reaching a population of more than 75,000 by 1945. The uranium processing facility, originally called the Clinton Engineer Works, was renamed the Oak Ridge National Laboratory.

The facility's first big project was the construction of the X-10 Graphite Reactor, which demonstrated that plutonium could be produced and extracted from enriched uranium. Shortly after the war, plutonium production for nuclear weapons moved elsewhere, and the laboratory was reorganized for peaceful nuclear research. This included the first production of medical isotopes for cancer treatment.

In 1950 the Oak Ridge School of Reactor Technology was established, and Oak Ridge researchers laid much of the groundwork for the development of reactors for power and propulsion in the following decade. Other reactors developed at Oak Ridge were designed for water desalination, radiation exposure research, as well as "fast" reactors that could breed their own fuel.

In later decades the emphasis on nuclear reactor research largely gave way to areas such as nuclear security, materials research (ceramics and alloys), health physics and fundamental physics, as well as supercomputing. The laboratory has hosted some of the world's fastest machines.

Union of Concerned Scientists

Website: http://www.ucsusa.org

The Union of Concerned Scientists (UCS) was founded in 1969 by a group of students and faculty at the Massachusetts Institute of Technology. This was a time when the continuing Cold War and the war in Vietnam were raising concerns about the militarization of science and technology.

In 1977 the UCS launched the "Scientists' Declaration on the Nuclear Arms Race," which called for an end to nuclear weapons tests as well as an end to the expanded deployment of nuclear weapons by both the United States and the Soviet Union. During the 1980s the UCS came out against Star Wars—the Strategic Defense Initiative—and the possible deployment of weapons in space.

The UCS also became involved in broader environmental concerns, supporting stronger pollution and fuel economy standards and the reduction of greenhouse gas emissions in an effort to reduce the threat of climate change.

With regard to nuclear power, the UCS has taken a generally critical stance, pointing out what its researchers see as flaws in the safety and security of nuclear installations and inadequate supervision by the Nuclear Regulatory Commission. An annual report is issued reviewing events and summarizing concerns.

U.S. Department of Energy, Office of Nuclear Energy

Website: http://www.ne.doe.gov

The Office of Nuclear Energy is a division of the U.S. Department of Energy (DOE). Its stated mission is "to advance nuclear power as a resource capable of meeting the Nation's energy, environmental, and national security needs by resolving technical, cost, safety, proliferation resistance, and security barriers through research, development, and demonstration as appropriate."

The office's research programs are aimed at four objectives:

- Develop technologies and other solutions that can improve the reliability, sustain the safety, and extend the life of current reactors.
- Develop improvements in the affordability of new reactors to enable nuclear energy to help meet the administration's energy security and climate change goals.
- Develop sustainable fuel cycles.
- Understand and minimize the risks of nuclear proliferation and terrorism.

In 2012 one of the office's new initiatives involved the development of small modular reactors, which advocates believe are safer and more versatile than older designs.

U.S. Nuclear Regulatory Commission

Website: http://www.nrc.gov

The Nuclear Regulatory Commission (NRC) is an independent agency of the U.S. government. The NRC was established by the Energy Reorganization Act of 1974. The NRC took on the regulatory agenda of the previous Atomic Energy

Commission (AEC), while another new agency—the Energy Research and Development Administration—assumed nonregulatory activities.

The NRC was charged with overseeing all aspects of U.S. commercial nuclear power, including licensing and safety monitoring reactors, and managing spent fuel (security, transportation, and storage). However, the NRC has been charged by critics with being lax and inconsistent in its regulation of nuclear processes and materials. More fundamentally, while the old AEC was dismantled because of concerns that it had been "captured" by the nuclear industry it was supposed to regulate, the NRC similarly has been criticized as having an inherent conflict between its dual mandate of regulating and promoting nuclear power.

Following the Fukushima nuclear meltdowns of 2011, the NRC recommended new standards that would push reactor operators to deal with catastrophes such as earthquakes and floods that might lead to a complete loss of power.

World Information Service on Energy

Website: http://www10.antenna.nl/wise/

The World Information Service on Energy grew out of antinuclear activists' desire in the 1970s to organize a newsletter to help them stay in touch with developing issues and coordinate their activities. In February 1978 nearly 200 people met in Amsterdam and began the publication. Today the organization has a number of "relay offices" in countries such as Argentina, Austria, the Czech Republic, India, Japan, and Russia.

Gradually activism began to play a more prominent role in the organization, for example, campaigns to close nuclear reactors in the Netherlands and keep nuclear power out of the Kyoto climate change process. The organization's website includes fact sheets and alerts on nuclear development and anti-nuclear actions.

World Nuclear Association

Website: http://www.world-nuclear.org

The World Nuclear Association (WNA) is the primary international organization promoting nuclear power and providing a variety of services to member corporations. Members are involved in every aspect of the industry, including the fuel cycle (uranium mining, enrichment, fuel fabrication, and transportation and disposal of spent fuel) as well as power plant operation.

Founded in 2001, the WNA is the successor to the Uranium Institute, which was established in 1975. The WNA also serves as a sort of umbrella group with affiliates that include professional organizations in the nuclear field as well as national and regional societies and industrial groups. The London-based group conducts an annual symposium.

The WNA website offers considerable background and policy information on all aspects of nuclear engineering and operations. The WNA also publishes the *World Nuclear News*.

This chapter presents illustrations, facts, statistics, and documents that provide background information about nuclear power and the issues surrounding nuclear power plant operation and the management of nuclear waste. Excerpts of longer documents are provided. Sources are indicated throughout so that readers can find additional information online.

Historical Background

President Truman on Civilian Control and Peaceful Use of Atomic Energy

In the wake of the shock caused by the use of the atomic bomb on Japan and the likelihood that the Soviet Union would also have such weapons, policy makers struggled to find a way to control nuclear weapons while opening possibilities for the peaceful use of nuclear power. President Truman proposed international control of nuclear weapons and began the process of creating a civilian nuclear program.

As President of the United States, I regard the continued control of all aspects of the atomic energy program, including research, development, and the custody of atomic weapons, as

Surfers pass in front of the San Onofre nuclear power plant in southern California. Plagued by problems with its steam generators, the plant closed in 2013. (AP Photo/Gregory Bull)

the proper functions of the civil authorities. Congress has recognized that the existence of this new weapon places a grave responsibility on the President as to its use in the event of a national emergency. There must, of course, be very close cooperation between the civilian Commission and the Military Establishment. Both the military authorities and the civilian Commission deserve high commendation for the joint efforts which they are putting forward to maintain our Nation's leadership in this vital work.

The Government of the United States holds in trust all our fissionable materials and production facilities. These are being used, on an increasing scale, to speed the discovery of applications of atomic energy to industry, agriculture, and public health. The Atomic Energy Commission reports that recent experiments hold out the promise of more efficient production on the farm and in the factory and of an increase of mechanical and human energy for doing the world's work. While this program is directed by an agency of the Government, the plants and laboratories are operated by leading industrial and research organizations through contracts with the Federal Government. I hope that, in due course, the Government will be able to permit greater participation by private industry in the development of atomic energy. When that time arrives, our industries and research organizations will be well prepared to carry forward the applications 'of atomic energy which will provide better living and better health for our people.

Secrecy is always distasteful to a free people. In scientific research, it is a handicap to productivity. But our need for security in an insecure world compels us, at the present time, to maintain a high order of secrecy in many of our atomic energy undertakings.

When the nations of the world are prepared to join with us in the international control of atomic energy, this requirement of secrecy will disappear. Our Government has sought, through its representatives on the United Nations Atomic Energy Commission, to find a common basis for understanding with the other member nations. However, the uncompromising refusal of the Soviet Union to participate in a workable control system has thus far obstructed progress.

The Atomic Energy Act has stood the test of 2 years of administration. There is no reason to question the sound basis on which it rests. In 2 years, the world has found no ready answers to the problem of war and peace. Atomic energy, therefore, remains a fearful instrument of destruction and a wonderful invitation to progress through peace.

Note: The Fourth Semiannual Report of the Atomic Energy Commission is printed in Senate Document 199 (80th Cong., 2d sess.).

Source: Harry S. Truman. "Statement by the President Reviewing Two Years of Experience with the Atomic Energy Act," July 24, 1948. Online by Gerhard Peters and John T. Woolley, American Presidency Project. http://www .presidency.ucsb.edu/ws/?pid=12966.

Domestic Development of Atomic Energy

In the Eisenhower administration the basic framework for the coming nuclear power industry was established. A number of steps were needed before nuclear materials could be used for reactors outside of direct government control and before private companies could begin to invest in building them.

What was only a hope and a distant goal in 1946—the beneficent use of atomic energy in human service—can soon be a reality. Before our scientists and engineers lie rich possibilities in the harnessing of atomic power. The Federal Government can pioneer in its development. But, in this undertaking, the enterprise, initiative and competitive spirit of individuals and groups within our free economy are needed to assure the greatest efficiency and progress at the least cost to the public.

Industry's interest in this field is already evident. In collaboration with the Atomic Energy Commission, a number of private corporations are now conducting studies, largely at their own expense, of the various reactor types which might be developed to produce economic power. There are indications that they would increase their efforts significantly if the way were open for private investment in such reactors. In amending the law to permit such investment, care must be taken to encourage the development of this new industry in a manner as nearly normal as possible, with careful regulation to protect the national security and the public health and safety. It is essential that this program so proceed that this new industry will develop self-reliance and self-sufficiency.

The creation of opportunities for broadened industrial participation may permit the Government to reduce its own reactor research and development after private industrial activity is well established. For the present, in addition to contributing toward the advancement of power reactor technology, the Government will continue to speed progress in the related technology of military propulsion reactors. The present complementary efforts of industry and Government will therefore continue, and industry should be encouraged by the enactment of appropriate legislation to assume a substantially more significant role. To this end, I recommend amendments to the Atomic Energy Act which would:

1. Relax statutory restrictions against ownership or lease of fissionable material and of facilities capable of producing fissionable material.

2. Permit private manufacture, ownership and operation of atomic reactors and related activities, subject to necessary safeguards and under licensing systems administered by the Atomic Energy Commission.

3. Authorize the Commission to establish minimum safety and security regulations to govern the use and possession of fissionable material.

4. Permit the Commission to supply licensees special materials and services needed in the initial stages of the new industry at prices estimated to compensate the Government adequately for the value of the materials and services and the expense to the Government in making them available.

5. Liberalize the patent provisions of the Atomic Energy Act, principally by expanding the area in which private patents can be obtained to include the production as well as utilization of fissionable material, while continuing for a limited period the authority to require a patent owner to license others to use an invention essential to the peacetime applications of atomic energy.

Until industrial participation in the utilization of atomic energy acquires a broader base, considerations of fairness require some mechanism to assure that the limited number of companies, which as government contractors now have access to the program, cannot build a patent monopoly which would exclude others desiring to enter the field. I hope that participation in the development of atomic power will have broadened sufficiently in the next five years to remove the need for such provisions.

In order to encourage the greatest possible progress in domestic application of atomic energy, flexibility is necessary in licensing and regulatory provisions of the legislation. Until further experience with this new industry has been gained, it would be unwise to try to anticipate by law all of the many problems that are certain to arise. Just as the basic Atomic Energy Act recognized by its own terms that it was experimental in a number of respects, so these amendments will be subject to continuing future change and refinement.

The destiny of all nations during the twentieth century will turn in large measure upon the nature and the pace of atomic energy development here and abroad. The revisions to the Atomic Energy Act herein recommended will help make it possible for American atomic energy development, public and private, to play a full and effective part in leading mankind into a new era of progress and peace.

> **Source**: Dwight D. Eisenhower: "Special Message to the Congress Recommending Amendments to the Atomic Energy Act," February 17, 1954. Online by Gerhard Peters and John T. Woolley, American Presidency Project. http://www.presidency.ucsb.edu/ws/?pid=10163.

President Kennedy and Planning for Development of Civilian Nuclear Power

As the 1960s began nuclear power was beginning to become a part of planning for future energy needs. The Kennedy administration thus undertook studies aimed toward facilitating the growth of domestic nuclear power as well as increasing its role in international cooperation.

[Letter to the Chairman, Atomic Energy Commission]

[Released March 20 1962. Dated March 17, 1962]

Dear Mr. Chairman:

The development of civilian nuclear power involves both national and international interests of the United States. At this time it is particularly important that our domestic needs and prospects for atomic power be thoroughly understood by both the Government and the growing atomic industry of this country which is participating significantly in the development of nuclear technology. Specifically we must extend our national energy resources base in order to promote our nation's economic growth.

Accordingly, the Atomic Energy Commission should take a new and hard look at the role of nuclear power in our economy in cooperation with the Department of the Interior, the federal Power Commission, other appropriate agencies, and private industry.

Your study should identify the objectives, scope, and content of a nuclear power development program in the light of the nation's prospective energy needs and resources and advances in alternate means for power generation.

It should recommend appropriate steps to assure the proper timing of development and construction of nuclear power projects, including the construction of necessary prototypes. There should, of course, be a continuation of the present fruitful cooperation between Government and industry—public utilities, private utilities and equipment manufacturers.

Upon completion of this study of domestic needs and resources, there should also be an evaluation of the extent to which our nuclear power program will further our international objectives in the peaceful uses of atomic energy.

The nuclear power plants scheduled to come into operation this year, together with those already in operation, should provide a wealth of engineering experience permitting realistic forecasts of the future of economically competitive nuclear power in this country.

As you are aware, two major related studies are now or will soon be under way. The study being conducted at my request by the National Academy of Sciences on the development and preservation of all our national resources will focus on the nation's longer-term energy needs and utilization of fuel resources. The other study to be launched soon by the federal Power Commission will determine the long range power requirements of the nation and will suggest the broad outline of possible programs of growth for all electric power companies-both private and public—to meet the great increase in power needs. Your study should be appropriately related to these investigations.

The extensive and vigorous atomic power development programs currently being undertaken by the Commission should, of course, be continued and, where appropriate, strengthened during the period of your study. I urge that your review be undertaken without delay and would hope that you could submit a report by September 1, 1962.

Sincerely,
JOHN F. KENNEDY

Note: The report "Civilian Nuclear Manpower" (67 pp., processed) was submitted to the President on November 20 by Dr. Glenn T. Seaborg, Chairman, Atomic Energy Commission.

Source: John F. Kennedy. "Letter to the Chairman, Atomic Energy Commission, on the Development of Civilian Nuclear Power," March 20, 1962. Online by Gerhard Peters and John T. Woolley, American Presidency Project. http://www.presidency.ucsb.edu/ws/?pid=8557.

President Carter's Nuclear Policy and a Turning Point in the 1970s?

The 1970s saw Three Mile Island and the rise of a loosely organized but influential antinuclear movement, and the future role of nuclear power was increasingly questioned. In 1977 President Jimmy Carter issued a statement reflecting his administration's review of nuclear issues, especially those involving production of nuclear fuel and potential proliferation.

Nuclear Power Policy Statement on Decisions Reached Following a Review.

April 7, 1977

There is no dilemma today more difficult to resolve than that connected with the use of nuclear power. Many countries see nuclear power as the only real opportunity, at least in this century, to reduce the dependence of their economic well-being on foreign oil—an energy source of uncertain availability, growing price, and ultimate exhaustion. The U.S., by contrast, has a major domestic energy source—coal—but its use is not without penalties, and our plans also call for the use of nuclear power as a share in our energy production.

The benefits of nuclear power are thus very real and practical. But a serious risk accompanies worldwide use of nuclear power—the risk that components of the nuclear power process will be turned to providing atomic weapons.

We took an important step in reducing the risk of expanding possession of atomic weapons through the nonproliferation treaty, whereby more than 100 nations have agreed not to develop such explosives. But we must go further. The U.S. is deeply concerned about the consequences for all nations of a further spread of nuclear weapons or explosive capabilities. We believe that these risks would be vastly increased by the further spread of sensitive technologies which entail direct access to plutonium, highly enriched uranium, or other weapons usable material. The question I have had under review from my first day in office is how can that be accomplished without forgoing the tangible benefits of nuclear power.

We are now completing an extremely thorough review of all the issues that bear on the use of nuclear power. We have concluded that the serious consequences of proliferation and direct implications for peace and security—as well as strong scientific and economic evidence—require:

—a major change in U.S. domestic nuclear energy policies and programs; and

—a concerted effort among all nations to find better answers to the problems and risks accompanying the increased use of nuclear power.

I am announcing today some of my decisions resulting from that review.

First, we will defer indefinitely the commercial reprocessing and recycling of the plutonium produced in the U.S. nuclear power programs. From our own experience, we have concluded that a viable and economic nuclear power program can be sustained without such reprocessing and recycling. The plant at Barnwell, South Carolina, will receive neither Federal encouragement nor funding for its completion as a reprocessing facility.

Second, we will restructure the U.S. breeder reactor program to give greater priority to alternative designs of the breeder and to defer the date when breeder reactors would be put into commercial use.

Third, we will redirect funding of U.S. nuclear research and development programs to accelerate our research into alternative nuclear fuel cycles which do not involve direct access to materials usable in nuclear weapons.

Fourth, we will increase U.S. production capacity for enriched uranium to provide adequate and timely supply of nuclear fuels for domestic and foreign needs.

Fifth, we will propose the necessary legislative steps to permit the U.S. to offer nuclear fuel supply contracts and guarantee delivery of such nuclear fuel to other countries.

Sixth, we will continue to embargo the export of equipment or technology that would permit uranium enrichment and chemical reprocessing.

Seventh, we will continue discussions with supplying and recipient countries alike, of a wide range of international approaches and frameworks that will permit all nations to achieve their energy objectives while reducing the spread of nuclear explosive capability. Among other things, we will explore the establishment of an international nuclear fuel cycle evaluation program aimed at developing alternative fuel cycles and a variety of international and U.S. measures to assure access to nuclear fuel supplies and spent fuel storage for nations sharing common nonproliferation objectives.

We will continue to consult very closely with a number of governments regarding the most desirable multilateral and bilateral arrangements for assuring that nuclear energy is creatively harnessed for peaceful economic purposes. Our intent is to develop wider international cooperation in regard to this vital issue through systematic and thorough international consultations.

Source: Jimmy Carter. "Nuclear Power Policy Statement on Decisions Reached Following a Review," April 7, 1977. Online by Gerhard Peters and John T. Woolley, American Presidency Project. http://www.presidency.ucsb.edu/ws/?pid=7316.

President Reagan's Attempt to Redirect Nuclear Policy

As in other policy matters, the Reagan administration took a more "pro-growth" stance. However, most of the developments anticipated in the following statement would not come to fruition.

A more abundant, affordable, and secure energy future for all Americans is a critical element of this administration's economic recovery program. While homeowners and business firms have shown remarkable ingenuity and resourcefulness in meeting their energy needs at lower cost through conservation, it is evident that sustained economic growth over the decades ahead will require additional energy supplies. This is particularly true of electricity, which will supply an increasing share of our energy.

If we are to meet this need for new energy supplies, we must move rapidly to eliminate unnecessary government barriers to efficient utilization of our abundant, economical resources of coal and uranium. It is equally vital that the utilities—investor-owned, public, and co-ops—be able to develop new generating capacity that will permit them to supply their customers at the lowest cost, be it coal, nuclear, hydro, or new technologies such as fuel cells.

One of the best potential sources of new electrical energy supplies in the coming decades is nuclear power. The United States has developed a strong technological base in the production of electricity from nuclear energy. Unfortunately, the

Federal Government has created a regulatory environment that is forcing many utilities to rule out nuclear power as a source of new generating capacity, even when their consumers may face unnecessarily high electric rates as a result. Nuclear power has become entangled in a morass of regulations that do not enhance safety but that do cause extensive licensing delays and economic uncertainty. Government has also failed in meeting its responsibility to work with industry to develop an acceptable system for commercial waste disposal, which has further hampered nuclear power development.

To correct present government deficiencies and to enable nuclear power to make its essential contribution to our future energy needs, I am announcing today a series of policy initiatives:

(1) I am directing the Secretary of Energy to give immediate priority attention to recommending improvements in the nuclear regulatory and licensing process. I anticipate that the Chairman of the Nuclear Regulatory Commission will take steps to facilitate the licensing of plants under construction and those awaiting licenses. Consistent with public health and safety, we must remove unnecessary obstacles to deployment of the current generation of nuclear power reactors. The time involved to proceed from the planning stage to an operating license for new nuclear power plants has more than doubled since the mid-1970's and is presently some 10-14 years. This process must be streamlined, with the objective of shortening the time involved to 6-8 years, as is typical in some other countries.

(2) I am directing that government agencies proceed with the demonstration of breeder reactor technology, including completion of the Clinch River Breeder Reactor. This is essential to ensure our preparedness for longer-term nuclear power needs.

(3) I am lifting the indefinite ban which previous administrations placed on commercial reprocessing activities in the United States. In addition, we will pursue consistent, long-term policies concerning reprocessing of spent fuel from

nuclear power reactors and eliminate regulatory impediments to commercial interest in this technology, while ensuring adequate safeguards.

It is important that the private sector take the lead in developing commercial reprocessing services. Thus, I am also requesting the Director of the Office of Science and Technology Policy, working with the Secretary of Energy, to undertake a study of the feasibility of obtaining economical plutonium supplies for the Department of Energy by means of a competitive procurement. By encouraging private firms to supply fuel for the breeder program at a cost that does not exceed that of government-produced plutonium, we may be able to provide a stable market for private sector reprocessing and simultaneously reduce the funding needs of the U.S. breeder demonstration program.

(4) I am instructing the Secretary of Energy, working closely with industry and State governments, to proceed swiftly toward deployment of means of storing and disposing of commercial, high-level radioactive waste. We must take steps now to accomplish this objective and demonstrate to the public that problems associated with management of nuclear waste can be resolved.

(5) I recognize that some of the problems besetting the nuclear option are of a deep-seated nature and may not be quickly resolved. Therefore, I am directing the Secretary of Energy and the Director of the Office of Science and Technology Policy to meet with representatives from the universities, private industry, and the utilities, and requesting them to report to me on the obstacles which stand in the way of increased use of nuclear energy and the steps needed to overcome them in order to assure the continued availability of nuclear power to meet America's future energy needs, not later than September 30, 1982.

Eliminating the regulatory problems that have burdened nuclear power will be of little use if the utility sector cannot raise the capital necessary to fund construction of new generating facilities. We have already taken significant steps to improve the climate for capital formation with the passage of my program for economic recovery. The tax bill contains substantial incentives designed to attract new capital into industry.

Safe commercial nuclear power can help meet America's future energy needs. The policies and actions that I am announcing today will permit a revitalization of the U.S. industry's efforts to develop nuclear power. In this way, native American genius, not arbitrary Federal policy, will be free to provide for our energy future.

Source: Ronald Reagan. "Statement Announcing a Series of Policy Initiatives on Nuclear Energy," October 8, 1981. Online by Gerhard Peters and John T. Woolley, American Presidency Project. http://www.presidency .ucsb.edu/ws/?pid=44353.

The Nuclear Industry

After more than 50 years nuclear power is a mature industry with plants operating in many countries. The United States is the world's largest producer of nuclear power, accounting for more than 30 percent of worldwide production. A number of factors will determine the future growth (or decline) of this industry, including:

- Whether concerns about nuclear plant safety and security can be resolved to the satisfaction of policy makers and the public, particularly in the wake of the 2011 Fukushima Daiichi disaster
- The immediate and total (life cycle) cost of nuclear power compared to fossil (oil, gas, and coal) or renewable (hydroelectric, wind, solar, etc.) energy sources
- The degree to which nuclear power may be perceived to be the only practicable immediate source for the shift from fossil fuels needed to combat global climate change

- The degree to which some combinations of conservation and increased use of renewable energy may be sufficient to accomplish climate goals

The following sections summarize aspects of the operation of the nuclear industry in the United States and around the world.

Nuclear and Other Forms of Energy

While nuclear power contributes a significant portion of the energy used in the United States, three fossil fuel sources—petroleum, natural gas, and coal—all have larger shares. Other sources such as hydroelectric and other renewables have smaller but growing shares. Nuclear's share of energy production is higher (about 19 percent) than its share of actual consumption, which was about

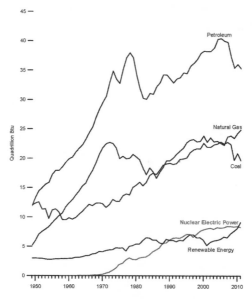

Figure 5.1   Primary Energy Consumption Estimates by Major Source, 1949–2011 (U.S. Energy Information Administration Annual Energy Review. September 2012. http://www.eia.gov/totalenergy/data/annual/pdf/aer.pdf)

8 percent in 2011. This discrepancy is due to the fact that nuclear energy is delivered in the form of electricity, which has a lower efficiency than energy directly derived from burning oil or natural gas.

Due to increased conservation and the improved efficiency of many devices, U.S. energy consumption per capita declined by 12 percent between 2000 and 2009. The EIA predicts a further decline of 8 percent between 2012 and 2040 (see Figure 5.1).

Source: U.S. Energy Information Administration Annual Energy Review. September 2012. http://www.eia.gov/totalenergy/data/annual/pdf/aer.pdf

Nuclear Power Around the World

The biggest producers of nuclear power (more than 100 TWh, or terawatt [trillion watt] hours) have been the United States, France, Russia, Japan, South Korea, and Germany. There is considerable variation in the role played by nuclear energy around the world and in how the share of energy generated by nuclear plants has changed over time. For example, Germany and Japan saw their nuclear share roughly halved between 2001 and 2011, while France has held steady, receiving about three quarters of its energy from nuclear power. The longer term effects of the 2011 Fukushima disaster on commitment to nuclear power remained unclear (see Table 5.1).

Growth of the U.S. Nuclear Industry

The number of operating nuclear plants in the United States has remained rather constant over the years, with no new plants coming online since 1990. However, there has been a rather steady increase in total output due to improvements in equipment and operating procedures (see Figure 5.2).

Projected New Nuclear Plants in the United States

As of of September 2013 the Nuclear Regulatory Commission had issued two new applications for building a total of four reactors,

Table 5.1 Nuclear Share Figures, 2002–2012

Country or Area	Nuclear Share (%)											Nuclear Electricity Production (TWh)	
	2002	2003	2004	2005	2006	2007	2008	2009	2010	2011	2012	2011	2012
Argentina	7.2	8.6	8.2	6.9	6.9	6.2	6.2	7.0	5.9	5.0	4.7	**5.9**	**5.9**
Armenia	40.5	35.5	38.8	42.7	42.0	43.5	39.4	45.0	39.4	33.2	26.6	**2.4**	**2.1**
Belgium	57.3	55.5	55.1	55.6	54.4	54.1	53.8	51.7	51.1	54.0	51.0	**45.9**	**38.5**
Brazil	4.0	3.6	3.0	2.5	3.3	2.8	3.1	3.0	3.1	3.2	3.1	**14.8**	**15.2**
Bulgaria	47.3	37.7	41.6	44.6	43.6	32.1	32.9	35.9	33.1	32.6	31.6	**15.3**	**14.9**
Canada	12.3	12.5	15.0	14.6	15.8	14.7	14.8	14.8	15.1	15.3	15.3	**88.3**	**89.1**
China:													
—Mainland	1.4	2.2	–	2.0	1.9	1.9	2.2	1.9	1.8	1.8	2.0	**82.6**	**92.7**
—Taiwan	22.9	21.5	–	–	19.5	19.3	17.1	20.7	19.3	19.0	18.4	**40.4**	**38.7**
Czech Rep	24.5	31.1	31.2	30.5	31.5	30.3	32.5	33.8	33.3	33.0	35.3	**26.7**	**28.6**
Finland	29.8	27.3	26.6	32.9	28.0	28.9	29.7	32.9	28.4	31.6	32.6	**22.3**	**22.1**
France	78.0	77.7	78.1	78.5	78.1	76.9	76.2	75.2	74.1	77.7	74.8	**423.5**	**407.4**
Germany	29.9	28.1	32.1	31.0	31.8	25.9	28.3	26.1	28.4	17.8	16.1	**102.3**	**94.1**
Hungary	36.1	32.7	33.8	37.2	37.7	36.8	37.2	43.0	42.1	43.2	45.9	**14.7**	**14.8**
India	3.7	3.3	2.8	2.8	2.6	2.5	2.0	2.2	2.9	3.7	3.6	**28.9**	**29.7**
Iran	0	0	0	0	0	0	0	0	0	0	0.6	**–**	**1.3**
Japan	34.3	25.0	29.3	29.3	30.0	27.5	24.9	28.9	29.2	18.1	2.1	**156.2**	**17.2**
Korea, S	38.6	40.0	37.9	44.7	38.6	35.3	35.6	34.8	32.2	34.6	30.4	**147.8**	**143.5**

(continued)

Table 5.1 *(continued)*

Country or Area	Nuclear Share (%)											Nuclear Electricity Production (TWh)	
	2002	2003	2004	2005	2006	2007	2008	2009	2010	2011	2012	2011	2012
Lithuania	80.1	79.9	72.1	69.6	72.3	64.4	72.9	76.2	0	0	0	0	0
Mexico	4.1	5.2	5.2	5.0	4.9	4.6	4.0	4.8	3.6	3.6	4.7	9.3	8.4
Netherlands	4.0	4.5	3.8	3.9	3.5	4.1	3.8	3.7	3.4	3.6	4.4	3.9	3.7
Pakistan	2.5	2.4	2.4	2.8	2.7	2.3	1.9	2.7	2.6	3.8	5.3	3.8	5.3
Romania	10.3	9.3	10.1	8.6	9.0	13.0	17.5	20.6	19.5	19.0	19.4	10.8	10.7
Russia	16.0	16.5	15.6	15.8	15.9	16.0	16.9	17.8	17.1	17.6	17.8	162.0	166.3
Slovakia	65.4	57.3	55.2	56.1	57.2	54.3	56.4	53.5	51.8	54.0	53.8	14.3	14.4
Slovenia	40.7	40.4	38.8	42.4	40.3	41.6	41.7	37.9	37.3	41.7	36.0	5.9	5.2
South Africa	5.9	6.0	6.6	5.5	4.4	5.5	5.3	4.8	5.2	5.2	5.1	12.9	12.4
Spain	25.8	23.6	22.9	19.6	19.8	17.4	18.3	17.5	20.1	19.5	20.5	55.1	58.7
Sweden	45.7	49.6	51.8	46.7	48.0	46.1	42.0	34.7	38.1	39.6	38.1	58.1	61.5
Switzerland	39.5	39.7	40.0	32.1	37.4	40.0	39.2	39.5	38.0	40.8	35.9	25.7	24.4
UK	22.4	23.7	19.4	19.9	18.4	15.1	13.5	17.9	15.7	17.8	18.1	62.7	64.0
Ukraine	45.7	45.9	51.1	48.5	47.5	48.1	47.4	48.6	48.1	47.2	46.2	84.9	84.9
USA	20.3	19.9	19.9	19.3	19.4	19.4	19.7	20.2	19.6	19.2	19.0	790.4	770.7
TOTAL												2518	2346

Note: – data not yet available

Source: World Nuclear Association. "Nuclear Share Figures, 2002–2012." May 2013. Available at http://www.world-nuclear.org/info/Facts-and-Figures/Nuclear-generation-by-country/. Used by permission.

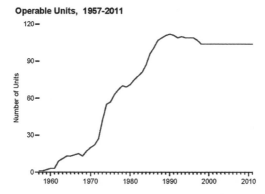

Figure 5.2 Operable Units, 1957–2011 (U.S. Energy Information Administration Annual Energy Review. September 2012. http://www.eia.gov/totalenergy/data/ annual/pdf/aer.pdf)

and nine applications involving 14 reactors were under review.Starting a licensing process does not mean that the facility will necessarily be built. The expanded accessibility of oil and

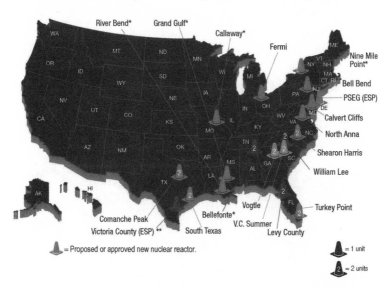

Figure 5.3 Locations of New Nuclear Power Reactor Applications (Nuclear Regulatory Commission, http://www.nrc.gov/reactors/new-reactors/col/new-reactor -map.html)

natural gas through new processes such as hydraulic "fracking" that lead to lower fossil fuel prices will have an as yet unknown impact on the economic viability of nuclear power (see Figure 5.3).

Where Are the U.S. Nuclear Plants?

The following map shows the distribution of nuclear power plants in the continental United States. The siting of nuclear plants involves somewhat contradictory needs. Siting plants closer to cities makes for more efficient distribution of the generated electricity. On the other hand, if a plant is near populated areas, the consequences of accidental release of radiation or other disasters will affect more people. Other siting considerations include the availability of cooling water (typically from a river). Coastal areas offer a source of water for emergency cooling but risk damage from flood waves (as was experienced in Fukushima, Japan, in 2011) or hurricanes. There are also seismic considerations in earthquake-prone areas such as California and Japan (see Figure 5.4).

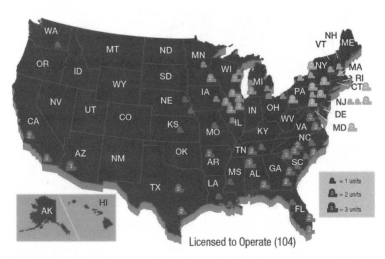

Licensed to Operate (104)

Figure 5.4 U.S. Operating Commercial Nuclear Power Reactors (Nuclear Regulatory Commission)

Basic Statistics on U.S. Nuclear Plants

The following commercial nuclear power plants were operating in the United States as of 2013. Additional information can be found by searching for the plant name online or through the website of the Nuclear Regulatory Commission (NRC), the Energy Information Administration (EIA), or the World Nuclear Association (an industry group) (see Table 5.2).

Table 5.2   U.S. Operating Nuclear Reactors

Reactor	State	Type	Net capacity (MWe)
Arkansas Nuclear One 1	Arkansas	PWR	836
Arkansas Nuclear One 2	Arkansas	PWR	992
Beaver Valley 1	Pennsylvania	PWR	892
Beaver Valley 2	Pennsylvania	PWR	914
Braidwood 1	Illinois	PWR	1,178
Braidwood 2	Illinois	PWR	1,152
Browns Ferry 1	Alabama	BWR	1,101
Browns Ferry 2	Alabama	BWR	1,104
Browns Ferry 3	Alabama	BWR	1,105
Brunswick 1	North Carolina	BWR	938
Brunswick 2	North Carolina	BWR	932
Byron 1	Illinois	PWR	1,164
Byron 2	Illinois	PWR	1,136
Callaway	Missouri	PWR	1,190
Calvert Cliffs 1	Maryland	PWR	866
Calvert Cliffs 2	Maryland	PWR	850
Catawba 1	South Carolina	PWR	1,129
Catawba 2	South Carolina	PWR	1,129
Clinton	Illinois	BWR	1,065
Columbia 2	Washington	BWR	1,153
Comanche Peak 1	Texas	PWR	1,205
Comanche Peak 2	Texas	PWR	1,195
Cooper	Nebraska	BWR	776
Davis Besse	Ohio	PWR	894
Diablo Canyon 1	California	PWR	1,122
Diablo Canyon 2	California	PWR	1,118
Donald C. Cook 1	Michigan	PWR	1,009

(continued)

Table 5.2 *(continued)*

Reactor	State	Type	Net capacity (MWe)
Donald C. Cook 2	Michigan	PWR	1,060
Dresden 2	Illinois	BWR	883
Dresden 3	Illinois	BWR	867
Duane Arnold	Iowa	BWR	601
Edwin I. Hatch 1	Georgia	BWR	876
Edwin I. Hatch 2	Georgia	BWR	883
Fermi 2	Michigan	BWR	1,085
Fort Calhoun	Nebraska	PWR	479
Ginna	New York	PWR	581
Grand Gulf 1	Mississippi	BWR	1,419
H.B. Robinson 2	South Carolina	PWR	741
Hope Creek 1	New Jersey	BWR	1,174
Indian Point 2	New York	PWR	1,020
Indian Point 3	New York	PWR	1,041
James A. Fitzpatrick	New York	BWR	848
Joseph M. Farley 1	Alabama	PWR	874
Joseph M. Farley 2	Alabama	PWR	860
La Salle 1	Illinois	BWR	1,137
La Salle 2	Illinois	BWR	1,140
Limerick 1	Pennsylvania	BWR	1,146
Limerick 2	Pennsylvania	BWR	1,150
McGuire 1	North Carolina	PWR	1,129
McGuire 2	North Carolina	PWR	1,129
Millstone 2	Connecticut	PWR	869
Millstone 3	Connecticut	PWR	1,233
Monticello	Minnesota	BWR	625
Nine Mile Point 1	New York	BWR	630
Nine Mile Point 2	New York	BWR	1,144
North Anna 1	Virginia	PWR	943
North Anna 2	Virginia	PWR	943
Oconee 1	South Carolina	PWR	846
Oconee 2	South Carolina	PWR	846
Oconee 3	South Carolina	PWR	846
Oyster Creek 1	New Jersey	BWR	615
Palisades	Michigan	PWR	782
Palo Verde 1	Arizona	PWR	1,311
Palo Verde 2	Arizona	PWR	1,314
Palo Verde 3	Arizona	PWR	1,312
Peach Bottom 2	Pennsylvania	BWR	1,125

(continued)

Table 5.2 *(continued)*

Reactor	State	Type	Net capacity (MWe)
Peach Bottom 3	Pennsylvania	BWR	1,125
Perry 1	Ohio	BWR	1,256
Pilgrim 1	Massachusetts	BWR	677
Point Beach 1	Wisconsin	PWR	591
Point Beach 2	Wisconsin	PWR	591
Prairie Island 1	Minnesota	PWR	521
Prairie Island 2	Minnesota	PWR	519
Quad Cities 1	Illinois	BWR	908
Quad Cities 2	Illinois	BWR	911
River Bend 1	Louisiana	BWR	975
Salem 1	New Jersey	PWR	1,168
Salem 2	New Jersey	PWR	1,158
Seabrook 1	New Hampshire	PWR	1,246
Sequoyah 1	Tennessee	PWR	1,152
Sequoyah 2	Tennessee	PWR	1,126
Shearon Harris 1	North Carolina	PWR	928
South Texas Project 1	Texas	PWR	1,280
South Texas Project 2	Texas	PWR	1,280
St. Lucie 1	Florida	PWR	981
St. Lucie 2	Florida	PWR	1135
Surry 1	Virginia	PWR	838
Surry 2	Virginia	PWR	838
Susquehanna 1	Pennsylvania	BWR	1,260
Susquehanna 2	Pennsylvania	BWR	1,260
Three Mile Island 1	Pennsylvania	PWR	805
Turkey Point 3	Florida	PWR	802
Turkey Point 4	Florida	PWR	802
V.C. Summer	South Carolina	PWR	971
Vermont Yankee 1	Vermont	BWR	604
Vogtle 1	Georgia	PWR	1,150
Vogtle 2	Georgia	PWR	1,152
Waterford 3	Louisiana	PWR	1,159
Watts Bar 1	Tennessee	PWR	1,123
Wolf Creek 1	Kansas	PWR	1,175
Total (102 units)			**99,098**

Reactor types: BWR (boiling water reactor) or PWR (pressurized water reactor)

Capacities are in megawatts electric (MWe).

Source: World Nuclear Association. "U.S. Operating Nuclear Reactors." December 2013. http://www.world-nuclear.org/info/Country-Profiles/Countries-T-Z/Appendices/Nuclear-Power-in-the-USA-Appendix-1—US-Operating-Nuclear-Reactors/. Used by permission.

Background: What Is Nuclear Energy?

Nuclear (or atomic) energy refers to energy that is derived from the forces that hold the nuclei (centers) of atoms together. This is different from chemical energy, which is derived from the electron bonds by which atoms combine to form molecules. Nuclear energy is much more highly concentrated than chemical energy, which is why a given mass of uranium-235 through fission can produce about 3 million times the energy available from burning the same mass in coal.

Nuclear Fission

Nuclear power is generated from a fission reaction that involves the splitting of atoms. Only atoms of certain isotopes (variants) of certain chemical elements are susceptible to fission. The most common isotopes used in nuclear fission reactions are uranium-235 and plutonium-239.

In a fission reaction some of the neutrons released by fission events go on to strike other atoms and cause additional fissions. If the rate at which this happens is sufficient, the reaction becomes self-sustaining. The fission rate can be controlled through the arrangement of fissionable fuel in the reactor and the use of materials that reflect, slow down, or absorb neutrons (moderators) (see Figure 5.5).

Nuclear Fusion

In a fusion reaction, atoms are combined rather than split. As with fission, a small amount of the mass of the original particles is converted to energy. In this example atoms of deuterium and tritium (isotopes of hydrogen with atomic masses of 2 and 3 respectively) combine to form a helium atom. They also release a neutron. Due to the intense pressure and heat involved in fusion, it is difficult to contain and sustain a fusion reaction, and despite decades of research, commercially viable fusion power remains an elusive goal (see Figure 5.6).

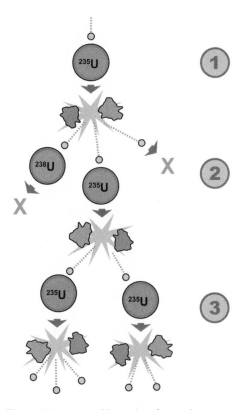

Figure 5.5   A possible nuclear fission chain reaction. 1. A uranium-235atom absorbs a neutron, and fissions into two new atoms (fission fragments), releasing three new neutrons and a large amount of binding energy. 2. One of those neutrons is absorbed by an atom of uranium-238, and does not continue the reaction. Another neutron leaves the system without being absorbed. However, one neutron does collide with an atom of uranium-235, which then fissions and releases two neutrons and more binding energy. 3. Both of those neutrons collide with uranium-235 atoms, each of which fissions and releases a few neutrons, which can then continue the reaction.

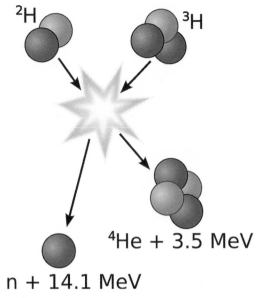

^{2}H ^{3}H

^{4}He + 3.5 MeV

n + 14.1 MeV

Figure 5.6   Nuclear Fusion

What Is Radiation, and What Does it Do?

The type of radiation released by radioactive materials (naturally or as a result of artificial nuclear reactions) is referred to as ionizing radiation, *which is radiation (energetic particles or waves) that has enough energy to ionize an atom by knocking an electron loose from it. Ionizing radiation should be distinguished from nonionizing radiation such as light or radio waves. Unlike nonionizing radiation, ionizing radiation has the potential to directly damage biological tissue.*

The three types of ionizing radiation released by radioactive materials differ in their energy and penetrating power. Alpha particles (helium nuclei) can be stopped by a sheet of paper. Beta particles (energetic electrons) can be blocked by a thin sheet of aluminum. However, gamma rays (a high energy form of electromagnetic radiation) require a dense material such as lead to block them (see Figure 5.7).

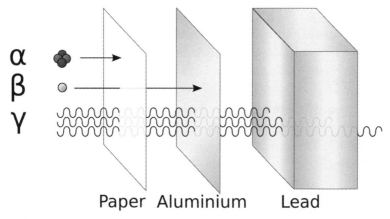

Figure 5.7   The three types of ionizing radiation released by radioactive materials differ in their energy and penetrating power. Alpha particles (helium nuclei) are stopped by a sheet of paper. Beta particles (energetic electrons) can be blocked by a thin sheet of aluminum. However gamma rays (a high energy form of electromagnetic radiation) require a dense material such as lead to block them. (Created by StannerEd)

Reactor Designs

While a nineteenth-century engineer would be amazed by the idea that atoms could be split to yield energy, and the core of a nuclear reactor would be a puzzle, he or she could quickly grasp the basic structure of the rest of a nuclear power plant. The idea of using heat to create steam pressure to do useful mechanical work goes back to the eighteenth century, while the steam turbine used in virtually all power reactors is a modernized version of an invention dating back to 1884. The main differences in reactor design have to do with the following:

- What fluid is used to cool the core and transfer the heat to where it can be put to work? This is usually water, although other fluids (such as molten sodium or a gas) can also be used.
- How is the steam generated from this heat? Is it generated directly from the first fluid, or from a second, separate fluid "loop"?

- What material is used as a moderator (a material that slows down neutrons from fission events to make them more likely to trigger additional fissions)? In many reactors water is both the coolant and moderator.

The most common types of power reactors are boiling water reactors (BWR) and pressurized water reactors (PWR). "Fast reactors" that generate energetic neutrons and can breed new fuel use liquid metal coolants because water would act as a moderator and slow down the neutrons too much.

Boiling Water Reactors

In a typical commercial boiling water reactor, (1) the core inside the reactor vessel creates heat; (2) a steam-water mixture is produced when very pure water (reactor coolant) moves upward through the core, absorbing heat; (3) the steam-water mixture leaves the top of the core and enters the two stages of moisture separation where water droplets are removed before the steam is allowed to enter the steam line; and (4) the steam line directs the steam to the main turbine, causing it to turn the turbine generator, which produces electricity. The unused steam is exhausted into the condenser, where it is condensed into water. The resulting water is pumped out of the condenser with a series of pumps, reheated, and pumped back to the reactor vessel. The reactor's core contains fuel assemblies that are cooled by water circulated using electrically powered pumps. These pumps and other operating systems in the plant receive their power from the electrical grid. If offsite power is lost, emergency cooling water is supplied by other pumps, which can be powered by onsite diesel generators. Other safety systems, such as the containment cooling system, also need electric power. Boiling water reactors contain between 370 and 800 fuel assemblies (see Figure 5.8).

Pressurized Water Reactors

In a typical commercial pressurized light water reactor, (1) the core inside the reactor vessel creates heat; (2) pressurized water in the

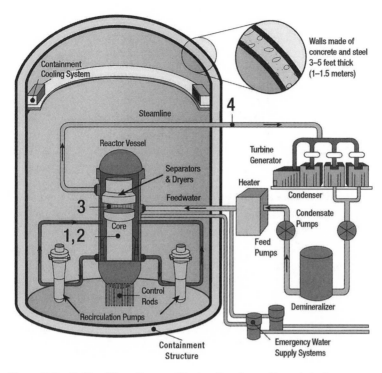

Walls made of
concrete and steel
3–5 feet thick
(1–1.5 meters)

Containment
Cooling System

Steamline

4

Reactor Vessel

Turbine
Generator

Separators
& Dryers

Heater

Feedwater

Condenser

3

Condensate
Pumps

Core

1,2

Feed
Pumps

Control
Rods

Demineralizer

Recirculation Pumps

Containment
Structure

Emergency Water
Supply Systems

Figure 5.8   Boiling-Water Reactor (Nuclear Regulatory Commission)

*primary coolant loop carries the heat to the steam generator; (3)
inside the steam generator, heat from the steam; and (4) the steam
line directs the steam to the main turbine, causing it to turn the
turbine generator, which produces electricity. The unused steam is
exhausted into the condenser, where it condensed into water. The
resulting water is pumped out of the condenser with a series of
pumps, reheated, and pumped back to the steam generators. The
reactor's core contains fuel assemblies that are cooled by water circulated using electrically powered pumps. These pumps and other
operating systems in the plant receive their power from the electrical
grid. If offsite power is lost, emergency cooling water is supplied by
other pumps, which can be powered by onsite diesel generators.
Other safety systems, such as the containment cooling system, also
need power. Pressurized water reactors contain between 150 and
200 fuel assemblies* (see Figure 5.9).

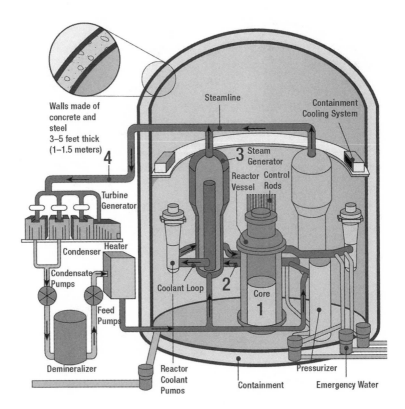

Figure 5.9   Pressurized-Water Reactor (Nuclear Regulatory Commission http://www.nrc.gov/reactors/pwrs.html)

Pebble Bed Reactors

The pebble bed reactor (PBR) is a relatively new reactor design characterized by how the fuel is fabricated and arranged. Instead of using rods filled with pellets as in conventional reactors, the fuel is encapsulated in tennis ball–sized spheres, or "pebbles," of graphite (which acts as a moderator) that contain the fuel particles. The tiny particles consist of fissionable material (such as uranium-235) coated with a silicon carbide ceramic. The pebbles in the core are cooled by a gas such as helium, nitrogen, or carbon dioxide, which does not react with the pebbles' coating.

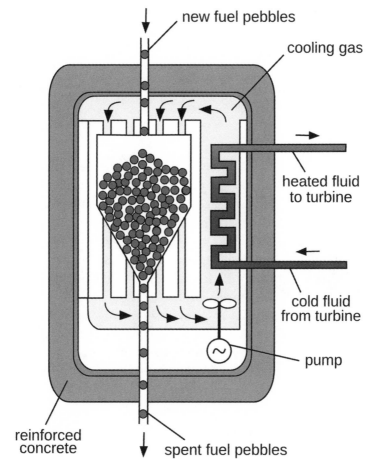

Figure 5.10   Pebble Bed Reactor

Pebble bed reactors are one type of VHTR (very high temperature reactor). Proponents of PBRs cite increased efficiency (because they run at high temperature) combined with increased safety because the reactor cools by natural circulation, and the pebbles help protect and isolate the radioactive fuel in case of accident (see Figure 5.10).

New Nuclear Plant Designs

In recent years the NRC has developed new regulations and procedures for reviewing and certifying new designs for nuclear reactors. The intent is to identify and resolve safety issues early in the development process and speed up the time needed to start building plants using a new design.

Background

The NRC has long sought standardization of nuclear power plant designs, and the enhanced safety and licensing reform that standardization could make possible. The Commission expects advanced reactors to be safer and use simplified, passive or other innovative means to accomplish their safety functions. The NRC's regulation (Part 52 to Title 10 of the Code of Federal Regulations) provides a predictable licensing process including certification of new nuclear plant designs. This process reflects decades of experience and research involving reactor design and operation. The design certification process provides for early public participation and resolution of safety issues prior to an application to construct a nuclear power plant.

Pre-Application Review Process

The NRC's "Statement of Policy for Regulation of Advanced Nuclear Power Plants," dated July 8, 1986, encourages early discussions, before a license application is submitted, between NRC and reactor designers to provide licensing guidance. In June 1988, the NRC issued NUREG-1226, "Development and Utilization of the NRC Policy Statement on the Regulation of Advanced Nuclear Power Plants." This document provides guidance on the implementation of the policy and describes the approach used by NRC in its review of advanced reactor design concepts.

In general, the NRC conducts pre-application reviews of advanced reactor designs to identify:

- major safety issues that could require Commission policy guidance to the staff,
- major technical issues that could be resolved under existing NRC regulations on policy, and
- research needed to resolve identified issues.

Design Certification Review Process

The review process for new reactor designs involves certifying standard reactor designs, independent of a specific site, through a rulemaking (Subpart B of Part 52). This rulemaking can certify a reactor design for 15 years. Design certification applicants must provide the technical information necessary to demonstrate compliance with the safety standards set forth in applicable NRC regulations (10 CFR Parts 20, 50, 73, and 100). Applicants must also provide information to close out unresolved and generic safety issues, as well as issues that arose after the Three Mile Island accident. The application must include a detailed analysis of the design's vulnerability to certain accidents or events, and inspections, tests, analyses, and acceptance criteria to verify the key design features. The NRC is considering a new rule that would require design certification applicants to assess their plant's level of built-in protection for avoiding or mitigating the effects of a large commercial aircraft impact, reducing the need for human intervention to protect [the] public.

Currently there are four certified reactor designs that can be referenced in an application for a combined license (COL) to build and operate a nuclear power plant. They are:

1. Advanced Boiling Water Reactor design by GE Nuclear Energy (May 1997);

Table 5.3   Design Certification Applications Currently Under Review

Design	Applicant
ABWR Design Certification Rule (DCR) Amendment	South Texas Project Nuclear Operating Company
Economic Simplified Boiling-Water Reactor (ESBWR)	GE-Hitachi Nuclear Energy
U.S. EPR	AREVA Nuclear Power
U.S. Advanced Pressurized-Water Reactor (US-APWR)	Mitsubishi Heavy Industries, Ltd.
ABWR Design Certification Renewal	Toshiba Corporation Power Systems Company
ABWR Design Certification Renewal	GE-Hitachi Nuclear Energy

2. System 80+ design by Westinghouse (formerly ABB-Combustion Engineering) (May 1997);

3. Advanced Passive (AP600) design by Westinghouse (December 1999); and

4. Advanced Passive (AP1000 design) by Westinghouse (January 2006).

The staff is currently reviewing the following design certification applications (see Table 5.3):

Design Descriptions. **ABWR**: The U.S. Advanced Boiling Water Reactor uses a single-cycle, forced circulation design with a rated power of 1,300 megawatts electric (MWe). The design incorporates features of the BWR designs in Europe, Japan, and the United States, and uses improved electronics, computer, turbine, and fuel technology. Improvements include the use of internal recirculation pumps, control rod drives that can be controlled by a screw mechanism rather than a step process, microprocessor-based digital control and logic systems, and digital safety systems. The design also includes safety enhancements such as protection against overpressurizing the containment, passive core debris flooding capability, an independent water makeup system, three emergency diesels, and a combustion turbine as an alternate power source.

AP600: The Advanced Passive 600 is a 600 MWe advanced pressurized water reactor that incorporates passive safety systems and simplified system designs. The passive systems use natural driving forces without active pumps, diesels, and other support systems after actuation. Use of redundant, non-safety related, active equipment and systems minimizes unnecessary use of safety-related systems.

AP1000: The Advanced Passive 1000 is a larger version of the previously approved AP600 design. This 1,100 MWe advanced pressurized water reactor incorporates passive safety systems and simplified system designs. It is similar to the AP600 design but uses a longer reactor vessel to accommodate longer fuel, and also includes larger steam generators and a larger pressurizer.

EPR: The Evolutionary Power Reactor (pictured at right) is a 1,600 MWe pressurized water reactor of evolutionary design. Design features include four 100% capacity trains of engineered safety features, a double-walled containment, and a "core catcher" for containment and cooling of core materials for severe accidents resulting in reactor vessel failure. The design does not rely on passive safety features. The first EPR is under construction at the Olkiluoto site in Finland, with another planned for the Flammanville site in France.

ESBWR: The Economic and Simplified Boiling Water Reactor is a 1,500 MWe, natural circulation boiling water reactor that incorporates passive safety features. This design is based on its predecessor, the 670 MWe Simplified BWR (SBWR) and also utilizes features of the certified ABWR. The ESBWR enhances natural circulation by using a taller vessel, a shorter core, and by reducing the flow restrictions.

The design utilizes an isolation condenser system for high-pressure water level control and decay heat removal during isolated conditions. After the automatic depressurization system operates, a gravity-driven cooling system provides low-pressure water level control. Containment cooling is provided by a passive system.

IRIS: The International Reactor Innovative and Secure is a pressurized light water cooled, medium-power (335 MWe) reactor that has been under development for several years by an international consortium. The IRIS design utilizes an integral reactor coolant system layout. The IRIS reactor vessel houses not only the nuclear fuel, control rods and neutron reflector, but also all the major reactor coolant system components including pumps, steam generators and pressurizer. The IRIS integral vessel is larger than a traditional PWR pressure vessel, but the size of the IRIS containment is a fraction of the size of corresponding loop reactors.

PBMR: The Pebble Bed Modular Reactor is a modular high-temperature gas reactor that uses helium as its coolant. PBMR design consists of eight reactor modules, 165 MWe per module, with capacity to store 10 years of spent fuel in the plant (there is additional storage capability in onsite concrete silos). The PBMR core is based on German high-temperature gas-cooled reactor technology and uses spherical graphite elements containing ceramic-coated fuel particles.

System 80+: This standard plant design uses a 1,300 MWe pressurized water reactor. It is based upon evolutionary improvements to the standard CE System 80 nuclear steam supply system and a balance-offplant design developed by Duke Power Co. The System 80+ design has safety systems that provide emergency core cooling, feedwater and decay heat removal. The new design also has a safety depressurization system for the reactor, a combustion turbine as an alternate AC power source, and an incontainment refueling water storage tank to enhance the safety and reliability of the reactor system.

Toshiba 4S: The Toshiba 4S reactor design has an output of about 10 MWe. The reactor has a compact core design, with steel-clad metal-alloy fuel. The core design does not require refueling over the 30-year lifetime of the plant. A three-loop configuration is used: primary system (sodium-cooled), an intermediate sodium loop between the radioactive primary system and the steam generators, and the water loop used to generate steam for the turbine. The basic layout is a "pool" configuration, with the pumps and intermediate heat exchanger inside the primary vessel.

US-APWR: The Mitsubishi Heavy Industry US-APWR design is an evolutionary 1,700 MWe pressurized water reactor currently being licensed and built in Japan. The design includes high-performance steam generators, a neutron reflector around the core to increase fuel economy, redundant core cooling systems and refueling water storage inside the containment building, and fully digital instrumentation and control systems.

> **Source:** Nuclear Regulatory Commission. "Backgrounder: New Nuclear Plant Designs," June 2008. http://www.nrc .gov/reading-rm/doc-collections/fact-sheets/new-nuc-plant -des-bg.pdf; "Design Certification Applications for New Reactors," April 2013. http://www.nrc.gov/reactors/new -reactors/design-cert.html

Fuel Cycle and Storage

To accurately assess the economics and environmental impact of nuclear power, it is necessary to look at the entire cycle of processing from the mining of natural uranium all the way to the ultimate storage or disposal of spent nuclear fuel. For example, while nuclear reactors themselves do not emit CO_2, uranium mining does have a "carbon footprint." Nevertheless, the carbon footprint of nuclear power is only a small fraction of that produced by burning fossil fuels, a fact that is emphasized by proponents of nuclear power as a way to reduce overall carbon emissions.

Stages of the Nuclear Fuel Cycle

The nuclear fuel cycle uses uranium in different chemical and physical forms. As illustrated in this section, this cycle typically includes the following stages (see Figure 5.11):

- Uranium recovery to extract (or mine) uranium ore, and concentrate (or mill) the ore to produce "yellowcake"
- Conversion of yellowcake into uranium hexafluoride (UF6)
- Enrichment to increase the concentration of uranium-235 (U235) in UF6
- Deconversion to reduce the hazards associated with the depleted uranium hexafluoride (DUF6), or "tailings," produced in earlier stages of the fuel cycle
- Fuel fabrication to convert enriched UF6 into fuel for nuclear reactors
- Use of the fuel in reactors (nuclear power, research, or naval propulsion)
- Interim storage of spent nuclear fuel

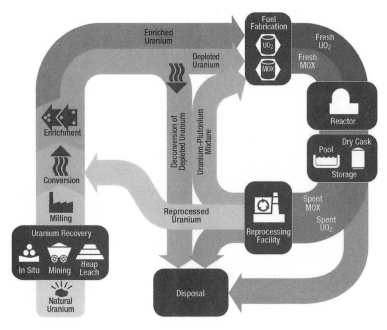

Figure 5.11 Stages of the Nuclear Fuel Cycle (Nuclear Regulatory Commission)

- Reprocessing (or recycling) of high-level waste (currently not done in the United States) [1]
- Final disposition (disposal) of high-level waste

Uranium Enrichment: Gas Centrifuge

Natural uranium has only a small proportion of the fissionable isotopes (mainly uranium-235) that are needed to create a chain reaction. Thus, physical or chemical processes must be used to raise the concentration of the fissionable isotopes sufficiently to provide reactor fuel. The two principle methods of uranium enrichment date back to the original work of the Manhattan Project during World War II.

The gas centrifuge uranium enrichment process uses a large number of rotating cylinders in series and parallel formations. Centrifuge machines are interconnected to form trains and cascades. In this process, uranium hexafluoride (UF6) gas is placed in a cylinder and rotated at a high speed. This rotation creates a strong centrifugal force so that the heavier gas molecules (containing U238) move toward the outside of the cylinder, and the lighter gas molecules (containing U235) collect closer to the center. The stream that is slightly enriched in U235 is withdrawn and fed into the next higher stage, while the slightly depleted stream is recycled back into the next lower stage. Significantly more U235 enrichment can be obtained from a single-unit gas centrifuge than from a single-unit gaseous diffusion stage. Figure 5.12 illustrates this gas centrifuge enrichment process.

One gas centrifuge commercial production plant, the URENCO USA facility owned by Louisiana Energy Services (LES), is currently operating in Eunice, New Mexico. A license was issued to AREVA Enrichment Services, LLC in October 2011 to construct and operate the Eagle Rock Enrichment Facility, a gas centrifuge commercial facility in Bonneville County, Idaho. The construction of the facility is currently on hold. USEC Inc., in Piketown, Ohio, was issued a license in

April 2007 to construct and operate the American Centrifuge Plant (ACP). Construction of the ACP began in 2007 but was placed on hold in 2009. In addition, USEC Inc. was granted a license in February 2004 for a demonstration and test gas centrifuge plant, the Lead Cascade facility, that is in operation.

The gas centrifuge process uses a large number of rotating cylinders in series and parallel configurations. Gas is introduced and rotated at high speed, concentrating the component of higher molecular weight toward the outer wall of the cylinder and the lower molecular weight component toward the center. The enriched and the depleted gases are removed by scoops.

Uranium Enrichment: Gaseous Diffusion

In a gaseous diffusion enrichment plant, uranium hexafluoride (UF6) gas is slowly fed into the plant's pipelines, where it is pumped through special filters called barriers or porous membranes. The holes in the barriers are so small that there is barely enough room for the UF6 gas molecules to pass through. The isotope enrichment occurs because the lighter UF6 gas molecules (with the U234 and U235 atoms) tend to diffuse faster through the barriers than the heavier UF6 gas molecules containing U238. One barrier isn't enough, though. It takes many hundreds of barriers, one after the other, before the UF6 gas contains enough U235 to be used in reactors. At the end of the process, the enriched UF6 gas is withdrawn from the pipelines and condensed back into a liquid that is poured into containers. The UF6 is then allowed to cool and solidify before it is transported to fuel fabrication facilities where it is turned into fuel assemblies for nuclear power plants. Figure 5.13 illustrates this gaseous diffusion enrichment process.

World Nuclear Reactors and Uranium Requirements

The following table gives an idea of how the demand for uranium varies around the world according to the number and capacity of nuclear power plants. It also includes proposed reactors that

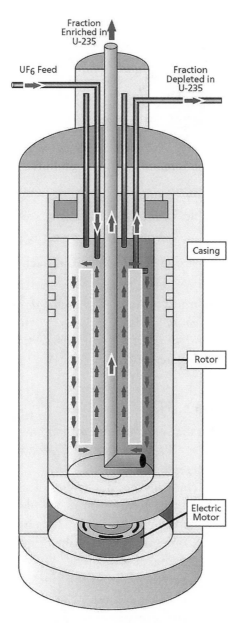

Figure   5.12 Uranium   Enrichment,   Gas Centrifuge (Nuclear Regulatory Commission)

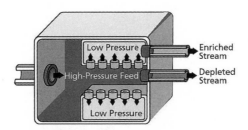

Figure 5.13 Uranium Enrichment, Gaseous Diffusion (Nuclear Regulatory Commission)

are expected to come online by 2030, but this is somewhat speculative (see Table 5.4).

Spent Nuclear Fuel and High-Level Radioactive Waste

Failure to complete the intended long-term nuclear waste depository in Yucca Mountain, Nevada, has resulted in spent fuel being stored indefinitely onsite at power plants. Additionally, when plants are decommissioned, the resulting high-level waste generally remains onsite (see Figure 5.14).

Decommissioning Nuclear Power Plants

The NRC regulates and monitors the decommissioning of nuclear plants. At the end of 2013 there were 17 nuclear power reactor sites undergoing decommissioning.

When a power company decides to close its nuclear power plant permanently, the facility must be decommissioned by safely removing it from service and reducing residual radioactivity to a level that permits release of the property and termination of the operating license. The Nuclear Regulatory Commission has strict rules governing nuclear power plant decommissioning, involving cleanup of radioactively contaminated plant systems and structures and removal of the radioactive fuel. These requirements protect workers and the public

Table 5.4   World Nuclear Power Reactors and Uranium Requirements (% e means percentage of total electricity generated)

Country (Click Name For Country Profile)	Nuclear Electricity Generation 2012		Reactors Operable December 2013		Reactors under Construction December 2013		Reactors Planned December 2013		Reactors Proposed December 2013		Uranium Required 2013
	billion kWh	% e	No.	MWe net	No.	MWe gross	No.	MWe gross	No.	MWe gross	tonnes U
Argentina	5.9	4.7	2	935	1	745	1	33	2	1400	212
Armenia	2.1	26.6	1	376	0	0	1	1060	0	0	86
Bangladesh	0	0	0	0	0	0	2	2000	0	0	0
Belarus	0	0	0	0	1	1200	1	1200	2	2400	0
Belgium	38.5	51.0	7	5943	0	0	0	0	0	0	1017
Brazil	15.2	3.1	2	1901	1	1405	0	0	4	4000	321
Bulgaria	14.9	31.6	2	1906	0	0	1	950	0	0	317
Canada	89.1	15.3	19	13553	0	0	2	1500	3	3800	1764
Chile	0	0	0	0	0	0	0	0	4	4400	0
China	92.7	2.0	18	14962	29	31610	59	64420	118	122000	6711
Czech Republic	28.6	35.3	6	3766	0	0	2	2400	1	1200	574
Egypt	0	0	0	0	0	0	1	1000	1	1000	0
Finland	22.1	32.6	4	2741	1	1700	0	0	2	3000	1127
France	407.4	74.8	58	63130	1	1720	1	1720	1	1100	9320
Germany	94.1	16.1	9	12003	0	0	0	0	0	0	1889
Hungary	14.8	45.9	4	1880	0	0	0	0	2	2200	357
India	29.7	3.6	21	5302	6	4300	18	15100	39	45000	1326

(continued)

Table 5.4 (continued)

Country (Click Name For Country Profile)	Nuclear Electricity Generation 2012		Reactors Operable December 2013		Reactors under Construction December 2013		Reactors Planned December 2013		Reactors Proposed December 2013		Uranium Required 2013
	billion kWh	% e	No.	MWe net	No.	MWe gross	No.	MWe gross	No.	MWe gross	tonnes U
Indonesia	0	0	0	0	0	0	2	2000	4	4000	0
Iran	1.3	0.6	1	915	0	0	1	1000	1	300	172
Israel	0	0	0	0	0	0	0	0	1	1200	0
Italy	0	0	0	0	0	0	0	0	10	17000	0
Japan	17.2	2.1	50	44396	3	3036	9	12947	3	4145	366
Jordan	0	0	0	0	0	0	1	1000	0	0	0
Kazakhstan	0	0	0	0	0	0	2	600	2	600	0
Korea DPR (North)	0	0	0	0	0	0	0	0	1	950	0
Korea RO (South)	143.5	30.4	23	20787	5	6870	6	8730	0	0	4218
Lithuania	0	0	0	0	0	0	1	1350	0	0	0
Malaysia	0	0	0	0	0	0	0	0	2	2000	0
Mexico	8.4	4.7	2	1600	0	0	0	0	2	2000	270
Netherlands	3.7	4.4	1	485	0	0	0	0	1	1000	103
Pakistan	5.3	5.3	3	725	2	680	0	0	2	2000	117
Poland	0	0	0	0	0	0	6	6000	0	0	0
Romania	10.6	19.4	2	1310	0	0	2	1310	1	655	177
Russia	166.3	17.8	33	24253	10	9160	31	32780	18	16000	5090

(continued)

238

Table 5.4 (continued)

Country (Click Name For Country Profile)	Nuclear Electricity Generation 2012 billion kWh	% e	Reactors Operable December 2013 No.	MWe net	Reactors under Construction December 2013 No.	MWe gross	Reactors Planned December 2013 No.	MWe gross	Reactors Proposed December 2013 No.	MWe gross	Uranium Required 2013 tonnes U
Saudi Arabia	0	0	0	0	0	0	0	0	16	17000	0
Slovakia	14.4	53.8	4	1816	2	942	0	0	1	1200	675
Slovenia	5.2	53.8	1	696	0	0	0	0	1	1000	137
South Africa	12.4	5.1	2	1830	0	0	0	0	6	9600	305
Spain	58.7	20.5	7	7002	0	0	0	0	0	0	1357
Sweden	61.5	38.1	10	9508	0	0	0	0	0	0	1505
Switzerland	24.4	35.9	5	3252	0	0	0	0	3	4000	521
Thailand	0	0	0	0	0	0	0	0	5	5000	0
Turkey	0	0	0	0	0	0	4	4800	4	4500	0
Ukraine	84.9	46.2	15	13168	0	0	2	1900	11	12000	2352
UAE	0	0	0	0	2	2800	2	2800	10	14400	0
United Kingdom	64.0	18.1	16	10038	0	0	4	6680	9	12000	1828
USA	770.7	19.0	100	98951	5	6018	7	8463	15	24000	19622
Vietnam	0	0	0	0	0	0	4	4000	6	6700	0
WORLD**	2346	c 11	434	374,057	71	74,886	173	187,740	314	354,750	64,978

Source: World Nuclear Association. "World Nuclear Power Reactors & Uranium Requirements." December 2013. Available at http://www.world-nuclear.org/info/Facts-and-Figures/World-Nuclear-Power-Reactors-and-Uranium-Requirements/. Used by permission.

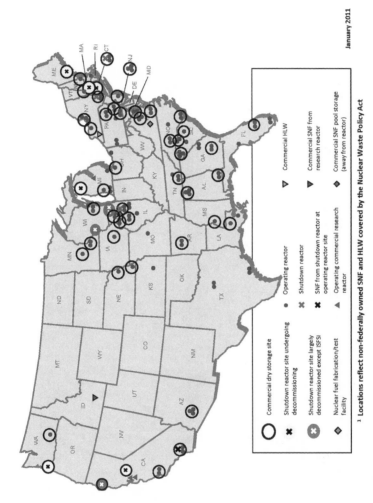

Figure 5.14 Locations of Spent Nuclear Fuel and High-Level Radioactive Waste[2] (U.S. Department of Energy, http://energy.gov/sites/prod/files/gcprod/documents/CurrentLocations_SNF_HLW_010111.pdf)

during the entire decommissioning process and the public after the license is terminated.

Discussion

Licensees may choose from three alternative decommissioning strategies: DECON, SAFSTOR, or ENTOMB.

- Under DECON (immediate dismantlement), soon after the nuclear facility closes, equipment, structures, and portions of the facility containing radioactive contaminants are removed or decontaminated to a level that permits release of the property and termination of the NRC license.
- Under SAFSTOR, often considered "delayed DECON," a nuclear facility is maintained and monitored in a condition that allows the radioactivity to decay; afterwards, it is dismantled and the property decontaminated.
- Under ENTOMB, radioactive contaminants are permanently encased on site in structurally sound material such as concrete and appropriately maintained and monitored until the radioactivity decays to a level permitting restricted release of the property. To date, no NRC-licensed facilities have requested this option.

The licensee may also choose to adopt a combination of the first two choices in which some portions of the facility are dismantled or decontaminated while other parts of the facility are left in SAFSTOR.

The decision may be based on factors besides radioactive decay such as availability of waste disposal sites.

To be acceptable, decommissioning must be completed within 60 years of the plant ceasing operations. A time beyond that would be considered only when necessary to protect public health and safety in accordance with NRC regulations.

Regulations

The requirements for decommissioning a nuclear power plant are set out in NRC regulations (Title 10 of the Code of Federal Regulations, Part 20 Subpart E, and Parts 50.75, 50.82, 51.53, and 51.95).

In August 1996, a revised rule went into effect that redefined the decommissioning process and required owners to provide the NRC with early notification of planned decommissioning activities.

The rule allows no major decommissioning activities to be undertaken until after certain information has been provided to the NRC and the public.

Decommissioning Funds

Each nuclear power plant licensee must report to the NRC every two years the status of its decommissioning funding for each reactor or share of a reactor that it owns. The report must estimate the minimum amount needed for decommissioning by using the formulas found in 10 CFR 50.75(c). Licensees may alternatively determine a site-specific funding estimate, provided that amount is greater than the generic decommissioning estimate. Although there are many factors that affect reactor decommissioning costs, generally they range from $300 million to $400 million.

Approximately 70 percent of licensees are authorized to accumulate decommissioning funds over the operating life of their plants. These owners—generally traditional, rate-regulated electric utilities or indirectly regulated generation companies—are not required today to have all of the funds needed for decommissioning. The remaining licensees must provide financial assurance through other methods such as prepaid decommissioning funds and/or a surety method or guarantee.

The staff performs an independent analysis of each of these reports to determine whether licensees are providing reasonable "decommissioning funding assurance" for radiological decommissioning of the reactor at the permanent termination of operation.

Before a nuclear power plant begins operations, the licensee must establish or obtain a financial mechanism—such as a trust fund or a guarantee from its parent company—to ensure that there will be sufficient money to pay for the ultimate decommissioning of the facility.

Public Involvement

Several opportunities are provided for public involvement during the decommissioning process. A public meeting is held in the vicinity of the facility after submittal of a post-shutdown decommissioning activities report (PSDAR) to the NRC. Another public meeting is held when NRC receives the license termination plan (LTP). An opportunity for a public hearing is provided prior to issuance of a license amendment approving the LTP or any other license amendment request. In addition, when NRC holds a meeting with the licensee, members of the public may observe the meeting (except when the discussion involves proprietary, sensitive, safeguards, or classified information).

Improving the Decommissioning Program

Several nuclear power plants completed decommissioning in the 1990s without a viable option for disposing of their spent nuclear fuel, because the federal government did not construct a geologic repository as planned. Accordingly, the NRC implemented regulations allowing licensees to sell off part of their land once it meets NRC release criteria, while maintaining a small parcel under license for storing the spent fuel. These stand-alone facilities, called "independent spent fuel storage installations" (ISFSIs), remain under license and NRC regulation. Licensees are responsible for security and for maintaining insurance and funding for eventual decommissioning.

As more facilities complete decommissioning, the NRC is implementing "lessons learned" in order to improve the program and focus on the prevention of future legacy sites. Applications for new reactors must now describe how design and operations will minimize contamination during the plant's operating life and facilitate eventual decommissioning. The agency is developing new regulations that will require plant operators to be more vigilant in preventing contamination during operations, and cleaning up and monitoring any contamination that does occur.

Phases of Decommissioning

The requirements for power reactor decommissioning activities may be divided into three phases: (1) initial activities; (2) major decommissioning and storage; and (3) license termination activities.

1) Initial Activities. When a nuclear power plant licensee shuts down the plant permanently, it must submit a written certification of permanent cessation of operations to the NRC within 30 days. When radioactive nuclear fuel is permanently removed from the reactor vessel, the owner must submit another written certification to the NRC, surrendering its authority to operate the reactor or load fuel into the reactor vessel. This eliminates the obligation to adhere to certain requirements needed only during reactor operation.

Within two years after submitting the certification of permanent closure, the licensee must submit a post-shutdown decommissioning activities report (PSDAR) to the NRC. This report provides a description of the planned decommissioning activities, along with a schedule for accomplishing them, and an estimate of the expected costs. The PSDAR must discuss the reasons for concluding that environmental impacts associated with the site-specific decommissioning activities have already been addressed in previous environmental analyses. Otherwise, the licensee must request a license amendment for approval of the activities and submit to the NRC a report on the additional impacts of decommissioning on the environment.

After receiving a PSDAR, the NRC publishes a notice of receipt in the Federal Register, makes the report available for public review and comment, and holds a public meeting in the vicinity of the plant to discuss the licensee's intentions.

2) Major Decommissioning Activities. Ninety days after the NRC receives the PSDAR, the owner can begin major decommissioning activities without specific NRC approval. These could include permanent removal of such major components as the reactor vessel, steam generators, large piping systems, pumps, and valves.

However, decommissioning activities conducted without specific prior NRC approval must not prevent release of the site for possible unrestricted use, result in there being no reasonable assurance that adequate funds will be available for decommissioning, or cause any significant environmental impact not previously reviewed.

If any decommissioning activity does not meet these terms, the licensee is required to submit a license amendment request, which would provide an opportunity for a public hearing.

Initially, the owner can use up to 3 percent of its funds set aside for decommissioning planning. An additional 20 percent can be used 90 days after submittal of the PSDAR. The remaining decommissioning trust funds are then available when the owner submits a detailed site-specific cost estimate to the NRC.

3) License Termination Activities. The owner is required to submit a LTP within two years of the expected license termination. The plan addresses each of the following: site characterization, identification of remaining site dismantlement activities, plans for site remediation, detailed plans for final radiation surveys for release of the site, method for demonstrating compliance with the radiological criteria for license termination, updated site-specific estimates of remaining decommissioning costs, and a supplement to the environmental report that describes any new information or significant environmental changes associated with the owner's proposed termination activities. Most plans envision releasing the site to the public for unrestricted use, meaning any residual radiation would be below NRC's limits of 25 millirem annual exposure and there would be no further regulatory controls by the NRC. Any plan proposing release of a site for restricted use must describe the site's end use, documentation on public consultation, institutional controls, and financial assurance needed to comply with the requirements for license termination for restricted release.

The LTP requires NRC approval of a license amendment. Before approval can be given, an opportunity for hearing is published and a public meeting is held near the plant site. The NRC uses a standard review plan (NUREG-1700, "Standard Review Plan for Evaluating Nuclear Power Reactor License Termination Plans)" to ensure high quality and uniformity of LTP reviews. The standard review plan is available to the public so that NRC's review process is understood clearly.

If the remaining dismantlement has been performed in accordance with the approved LTP and the termination survey demonstrates that the facility and site are suitable for release, the NRC issues a letter terminating the operating license.

More information on reactor decommissioning is available in responses to frequently asked questions at: http://www.nrc.gov/about-nrc/regulatory/decommissioning/faq.html.

Source: Nuclear Regulatory Commission. "Fact Sheet: Decommissioning Nuclear Power Plants," April 2011. http://www.nrc.gov/reading-rm/doc-collections/fact-sheets/decommissioning.pdf

Regulation, Safety and Risks

Many factors influence the safety of nuclear power plants. These include design features but also regulations, continuing safety assessment, and plans and procedures for dealing with a variety of accidents or emergencies.

Nuclear Regulatory Commission

The Nuclear Regulatory Commission is the agency of the U.S. government responsible for licensing and regulating nuclear power

How We Regulate

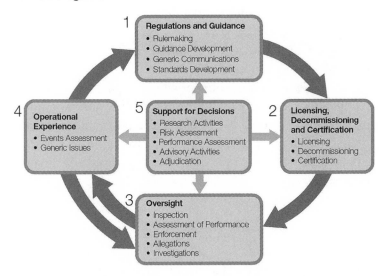

Figure 5.15   This diagram gives an overview of the NRC's regulatory process, which has five main components: (1) developing regulations and guidance for our applicants and licensees, (2) licensing or certifying applicants to use nuclear materials or operate nuclear facilities or decommissioning that permits license termination, (3) overseeing licensee operations and facilities to ensure that licensees comply with safety requirements, (4) evaluating operational experience at licensed facilities or involving licensed activities, and (5) conducting research, holding hearings to address the concerns of parties affected by agency decisions, and obtaining independent reviews to support our regulatory decisions. The NRC also strives to improve its processes in these five areas through risk-informed and performance-based regulation. (Nuclear Regulatory Commission. Available at http://www.nrc.gov/about-nrc/regulatory.html)

plants as well as nuclear fuel production and waste transport and disposal (see Figure 5.15).

The following activities are key components of our regulatory program.

Regulations and Guidance

Rulemaking - developing and amending regulations that licensees must meet to obtain or retain a license or certificate to use nuclear materials or operate a nuclear facility

Guidance Development - developing and revising guidance documents, such as regulatory guides, standard review plans, and NRC's Inspection Manual to aid licensees in meeting safety requirements

Generic Communications - sending applicants and licensees information about events or requests for information about regulatory requirements, some of which require response

Standards Development - working with industry standards organizations to develop consensus standards associated with systems, equipment, or materials used by the nuclear industry so that these standards may be referenced in NRC regulations or guidance

Licensing, Decommissioning, and Certification

Licensing - authorizing an applicant to use or transport nuclear materials or to operate a nuclear facility (includes new licenses, renewals, amendments, transfers and related Topical Reports)

Decommissioning - removing a nuclear facility safely from service and reducing residual radioactivity to a level that permits termination of license

Certification - authorizing an applicant to manufacture spent fuel casks, transportation packages for nuclear materials, and sealed sources and devices and authorizing an applicant to operate a gaseous diffusion plant

Oversight

Inspection - verifying that a licensee's activities are properly conducted to ensure safe operations in accordance with NRC's regulations

Enforcement - issuing sanctions to licensees who violate NRC regulations

Assessment of Performance (for operating facilities) - reviewing inspection findings, together with objective performance indicators, to assess the performance of nuclear facilities and determine appropriate agency action

Allegations- responding to reports of wrongdoing by NRC licensees, applicants for licenses, or licensee contractors or vendors

Investigations - investigating wrongdoing by NRC licensees

Operational Experience

Events Assessment - daily review and long term trend analysis of accidents and other reportable incidents to determine the appropriate regulatory response

Generic Issues - identifying and resolving safety issues that affect more than one licensed facility

Support for Decisions

Research Activities - experiments, technical studies, and analyses to help the NRC make realistic decisions, assess the safety-significance of potential technical issues, and prepare the agency for the future by evaluating potential safety issues involving new designs and technologies

Risk Assessment - use of risk analysis methods and performance insight to support decision making throughout the regulatory process

Performance Assessment (for waste disposal and decommissioning) - evaluating potential releases of radioactivity into the environment, and assessing the resultant radiological doses to demonstrate whether a disposal facility has met its performance objectives

Advisory Activities - review and assessment of regulatory proposals by independent advisory bodies reporting to or chartered by the Commission

CRGR Reviews - review and assessment of proposed generic backfits by the Committee To Review Generic Requirements (CRGR), an independent advisory committee chartered by the NRC's Executive Director for Operations

Adjudication - listening to concerns of parties affected by licensing or enforcement actions in a legal setting

Source: "How We Regulate." Nuclear Regulatory Commission. Available at http://www.nrc.gov/about-nrc/regulatory.html

International Nuclear Events Scale

The following scale attempts to rate nuclear "events" according to the intensity and scale of their effects. Both Chernobyl (1986) and Fukushima (2011) have been classified as level 7, Major Accident. Three Mile Island (1979) is rated as level 5. Recent criticism that the scale does not accurately reflect the real impact of disasters makes it likely it will be revised (see Figure 5.16).

Chernobyl Contamination Map

This map gives an indication of the spread of cesium-131, a radioisotope associated with increased incidence of thyroid cancer around the site of the Chernobyl nuclear plant following the disastrous meltdown and explosion in 1986. It has proven difficult to accurately measure the extent to which increases in cancer can be attributed to radioactive contamination as opposed to artifacts of the measuring process (see Figure 5.17).

INES events are
classified on the scale at
7–levels. Levels 1–3 are
called "incidents" and
Levels 4–7 "accidents."
The scale is designed so
that the severity of an
event is about 10 times
greater for each increase
in level on the scale.
Events without safety
significance are called
"deviations" and are
classified as Below Scale
or at Level 0.

Figure 5.16   The International Nuclear and Radiological Event Scale
(Nuclear Regulatory Commission)

NRC Recommendations in Response to the Fukushima Event

On July 12, 2011, shortly after the Fukushima Daiichi disaster, an NRC task force issued a report that included "enhancements to safety and emergency preparedness" for U.S. reactors. The recommendations were subsequently prioritized into three tiers.

Tier 1 Recommendations

The first tier consists of those actions which the staff determined should be started without unnecessary delay and for which sufficient resource flexibility, including availability of critical skill sets, exists.

- Seismic and flood hazard reevaluations (Recommendation 2.1)
- Seismic and flood walkdowns (Recommendation 2.3)
- Station blackout regulatory actions (Recommendation 4.1)
- Mitigating strategies for beyond design basis events (Recommendation 4.2)

Figure 5.17   Chernobyl Contamination Map (CIA Factbook)

- Reliable hardened vents for Mark I and Mark II containments (Recommendation 5.1)
- Spent fuel pool instrumentation (Recommendation 7.1)
- Strengthening and integration of emergency operating procedures, severe accident management guidelines, and extensive damage mitigation guidelines (Recommendation 8)
- Emergency preparedness regulatory actions (staffing and communications) (Recommendation 9.3)

Tier 2 Recommendations

The second tier consists of those actions which could not be initiated in the near term due to factors that include the need for further technical assessment and alignment, dependence on Tier 1 issues, or availability of critical skill sets. These actions do not require long-term study and can be initiated when sufficient technical information and applicable resources become available.

- Spent fuel pool makeup capability (Recommendation 7.2, 7.3, 7.4, and 7.5)
- Emergency preparedness regulatory actions (Recommendation 9.3)
- Other External Hazards Reevaluation (tornados, hurricanes, drought, etc.) (additional Issue)

Tier 3 Recommendations

The third tier consists of those actions that require further staff study to support a regulatory action, have an associated shorter-term action that needs to be completed to inform the longer-term action, are dependent on the availability of critical skill sets, or are dependent on the resolution of NTTF Recommendation 1. The staff has focused its initial efforts on developing the schedules, milestones, and resources associated with Tier 1 and Tier 2 activities. Once the staff has completed its evaluation of the resource impacts of the Tier 1 and Tier 2 recommendations, it will be able to more accurately address the Tier 3 recommendations.

- Ten-year confirmation of seismic and flooding hazards (dependent on Recommendation 2.1) (Recommendation 2.2)
- Potential enhancements to the capability to prevent or mitigate seismically-induced fires and floods (long-term evaluation) (Recommendation 3)
- Reliable hardened vents for other containment designs (long-term evaluation) (Recommendation 5.2)

- Hydrogen control and mitigation inside containment or in other buildings (longterm evaluation) (Recommendation 6)
- Emergency preparedness enhancements for prolonged station blackout and multiunit events (dependent on availability of critical skill sets) (Recommendation 9.1/9.2)
- Emergency Response Data System capability (related to long-term evaluation Recommendation 10) (Recommendation 9.3)
- Additional emergency preparedness topics for prolonged station blackout and multiunit events (long-term evaluation) (Recommendation 10)
- Emergency preparedness topics for decision-making, radiation monitoring, and public education (longterm evaluation) (Recommendation 11)
- Reactor Oversight Process modifications to reflect the recommended defense-indepth framework (dependent on Recommendation 1) (Recommendation 12.1)
- Staff training on severe accidents and resident inspector training on severe accident management guidelines (dependent on Recommendation 8) (Recommendation 12.2)
- Basis of emergency planning zone size (additional issue)
- Prestaging of potassium iodide beyond 10 miles (additional issue)
- Transfer of spent fuel to dry cask storage (additional issue)

Source: Nuclear Regulatory Commission. "NRC Japan Lessons Learned." An updated and more interactive version is available at http://www.nrc.gov/reactors/operating/ops-experience/japan-info.html

Union of Concerned Scientists: Nuclear Power Safety and Security Recommendations

The Union of Concerned Scientists, which has long taken a generally critical attitude toward the safety and future of nuclear power,

*issued a report urging stronger recommendations and require-
ments for power plant operation following the 2011 Fukushima
disaster.*

The recent events in Japan remind us that while the likelihood of a nuclear power plant accident is low, its potential consequences are grave. And an accident like Fukushima could
happen here. An equipment malfunction, fire, human error,
natural disaster or terrorist attack could—separately or in combination—lead to a nuclear crisis.

The U.S. will continue to obtain a significant portion of its
electricity from nuclear power for many years to come, regardless of how rapidly energy efficiency measures and other sources
of electricity are deployed.

Given this reality, the United States must take concrete steps
now to address serious shortcomings in nuclear plant safety and
security that have been evident for years. No technology can be
made perfectly safe, but the United States can and must do
more to guard against accidents as well as the threat of terrorist
attacks on reactors and spent fuel pools.

The report outlines and explains 23 specific recommendations, listed below.

* = Key recommendation that the NRC should make a top
priority.

Preventing and Mitigating the Effects of Severe Accidents

* **Extend the scope of regulations** to include the prevention
and mitigation of severe accidents.

Require reactor owners to develop and test emergency procedures for situations when no AC or DC power is available for
an extended period.

* **Modify emergency planning requirements** to ensure
that everyone at significant risk from a severe accident—not just
the people within the arbitrary 10-mile planning zone—is
protected.

Improving the Safety and Security of Spent Fuel

* The NRC should require plant owners to **move spent fuel at reactor sites from storage pools to dry casks** when it has cooled enough to do so.

The NRC should require reactor owners to improve the security of existing dry cask storage facilities.

The NRC should require plant owners to significantly improve emergency procedures and operator training for spent fuel pool accidents

Making Existing Reactors Safer

* The NRC should **enforce its fire protection regulations** and compel the owners of more than three dozen reactors to comply with regulations they currently violate.

* The NRC should **establish timeliness goals for resolving safety issues** while continuing to meet its timeliness goals for business-related requests from reactor owners.

The NRC should treat generic and unique safety issues alike. Until a generic issue is resolved, the NRC should account for it as a potential risk factor in its safety analyses and decision making related to all affected reactors.

The NRC should require plant owners to use multiple inspection techniques to ensure detection of any degradation in aging, high-risk equipment.

The NRC should require plant owners to periodically inspect equipment outside the scope of normal inspections, both to determine whether that scope is appropriate and to detect problems before safety margins are compromised.

The NRC should revise its regulations for the licensing of "high burn-up" fuel to ensure public safety, and restrict how this fuel is used until the revisions are complete.

The U.S. government should prohibit the use of plutonium-bearing mixed-oxide (MOX) fuel in reactors, and end the program to produce MOX fuel from excess weapons plutonium.

Ensuring the Continued Safety of Reactors with Renewed Licenses

Before granting a license renewal, the NRC should review all differences between current regulations and any past decisions specific to the aging reactor, to confirm that these differences will not compromise public safety going forward.

Making Existing Reactors More Secure against Terrorist Attacks

* The NRC should **revise its assumptions about terrorists' capabilities** to ensure nuclear plants are adequately protected against credible threats, and these assumptions should be reviewed by U.S. intelligence agencies.

The NRC should modify the way it judges force-on-force security exercises by assessing a plant's "margin to failure," rather than whether the plant merely passes or fails.

The U.S. government should establish a program for licensing private security guards that would require successful completion of a federally supervised training course and periodic recertification.

Making New Reactors More Secure against Terrorist Attacks

* The NRC should **require new reactor designs to be safer than existing reactors**.

The NRC should require new reactor designs to be more secure against land- and water-based terrorist attacks.

Improving the NRC's Cost-Benefit and Risk-Informed Analyses

* The NRC should **increase the value it assigns to a human life** in its cost-benefit analyses so the value is consistent with other government agencies.

The NRC should require plant owners to calculate the risk of fuel damage in spent fuel pools as well as reactor cores in all safety analyses.

The NRC should not make decisions about reactor safety using probabilistic risk assessments (PRAs) until it has corrected its flawed application of this tool.

Ensuring Public Participation

The NRC should fully restore the public's right to obtain information and question witnesses in hearings about changes to existing power plant licenses and applications for new licenses.

Source: Union of Concerned Scientists. "Nuclear Power Safety & Security Recommendations," July 2011. http://www.ucsusa.org/nuclear_power/nuclear_power_risk/safety/ucs-nuclear-safety-recommendations.html. Used by permission of the Union of Concerned Scientists, *www.ucsusa.org*.

NRC Response to Emergency Events

The NRC has developed a flexible response system to deal with nuclear emergencies that may require measures such as sheltering or evacuation and that require cooperation with plant officials and local authorities.

In the unlikely event of a serious emergency involving an NRC-license facility or material, the agency is prepared to respond immediately. Trained personnel continuously monitor licensee activities and are available to take information about a variety of threats from other federal agencies. In addition, specially trained responders in a variety of disciplines are always on call and able to respond quickly. Equipment, policies, and procedures for these response activities are regularly tested, re-evaluated, and updated so that the agency is ready at all times.

If a significant incident occurs, the NRC activates its Headquarters Operations Center (HOC) and one or more of its four Regional Incident Response Centers (IRCs). Specially trained and qualified personnel work in the HOC at all times. They take emergency information from a licensee and immediately notify key NRC managers and staff. These managers and staff are trained as responders in their areas of expertise and assemble in the HOC and IRC to support NRC response activities. The responders' activities are defined by the agency's incident response program.

Incident Response Program

The NRC incident response program uses a flexible system to tailor its response to the significance of an event. In the different modes of the system, NRC responders' activities change in order to best support the event. For example, in an event involving an NRC-licensed facility, the agency might decide to move from its "normal" response mode to "monitoring mode". In this level of response, key regional experts staff the appropriate IRC to respond to the event. If necessary, the NRC could then enter "activation mode". In this mode, the necessary safety, security, and preparedness specialists report to the HOC. The final emergency response mode is called "expanded activation". It is entered when an incident's severity or uncertainty warrants sending a team of NRC experts directly to the site of the event. Once the team arrives at the site and assesses the situation, oversight of the incident may be transferred from Headquarters to the Site Team.

When the NRC incident response program is activated to any mode beyond normal, the agency notifies skilled and trained responders who assemble and begin working directly with various counterparts. In the example of an event involving an NRC-licensed facility, some responders will work directly with the nuclear power plant operators to assess the condition of the plant and local officials.

These protective actions may include sheltering, evacuation, or the use of potassium iodide where appropriate. Other NRC responders will relay information about the incident to the media, states, local governments, tribal entities, other federal agencies, Congress, the White House, and international governments.

Although the licensee has primary responsibility to stabilize their facility or material and return it to a safe condition, the NRC Chairman has the authority to intervene and direct the licensee's on-site response if necessary to protect public health and safety and the environment.

Equipment, policies, and procedures for response activities are regularly tested, reevaluated, and updated so that the agency is ready at all times. The NRC tests itself many times each year with drills and exercises that mimic safety or security incidents and test the response plans of the agency and licensed facilities. In addition to full-scale exercises, the HOC and IRCs are periodically activated throughout the year for small emergencies or potential emergencies.

Licensee Response

NRC regulations require licensees to have plans for responding to incidents, protecting against radiological releases, and reducing the impacts of incidents. The NRC reviews these plans on a regular basis and tests them through exercises.

If a significant incident or emergency occurs, licensees are required to take immediate actions to ensure safety and security. They must also provide timely notifications to the NRC and state and local government authorities, and recommend how to protect the public from potential consequences.

Based on NRC regulations, licensees classify incidents according to the plant conditions and the level of risk to the public. Nuclear power plants, for example, use four emergency classifications:

• **Notification of Unusual Event**- Under this category, events are in process or have occurred that indicate a potential decline in the level of safety of the plant. No release of radioactive material requiring offsite response or monitoring is expected at that time.

- **Alert**- If an alert is declared, events are in process or have occurred that involve an actual or potentially substantial decline in the level of plant safety. However, any release of radioactive material is expected to be only a small fraction of the Environmental Protection Agency (EPA) protective action guidelines.

- **Site Area Emergency**- A site area emergency involves events in progress or which have occurred that result in an actual or likely a major failure of the plant's ability to protect the public. Any releases of radioactive material are not expected to exceed the EPA guidelines except near the site boundary.

- **General Emergency**- A general emergency involves actual or imminent severe damage or melting of radioactive fuel in the reactor core with the potential for loss of containment integrity. Radioactive releases during a general emergency can be expected to exceed the EPA guidelines beyond the immediate site area.

State and Local Government Response

State governments, and in some locations, local and/or tribal governments, develop and implement emergency plans for incidents involving an NRC-licensed facility or material. Although the licensee is the primary party responsible for what occurs onsite, state and local governments are responsible for protecting life, property, and the environment offsite.

Through drills and exercises, state and local governments work closely with the Federal Emergency Management Agency, and when appropriate, the NRC, to ensure that their plans and procedures will protect their community's health and safety.

During an emergency incident, the NRC communicates directly with state and local governments to share information. The NRC may also offer technical advice and assistance if requested.

Federal Response

The NRC works within the National Response Framework to respond to events. The framework guides the nation in how to respond to complex events that may involve a variety of agencies and hazards.

Under this framework, the NRC retains its independent authority and ability to respond to emergencies that involve NRC-licensed facilities or materials. The NRC coordinates the federal technical response to an incident that involves one of its licensees.

The NRC may request the support of the Department of Homeland Security (DHS) in responding to an emergency at an NRC-licensed facility or involving NRC-licensed materials. DHS may lead and manage the overall federal response to an event, according to Homeland Security Presidential Directive-5. In this case, the NRC would perform an important role in providing technical expertise and helping share information among the various organizations and licensees.

> **Source**: Nuclear Regulatory Commission, "Fact Sheet on Incident Response," March 2012. http://www.nrc.gov/reading-rm/doc-collections/fact-sheets/fs-incident-response.html

What Do I Do in a Nuclear Emergency?

If you live or work near a nuclear power plant, it would be good to be aware of these instructions from the NRC. Of course, major or large-scale nuclear disasters would be widely covered by a variety of media—but remember that initial reports are often incomplete or inaccurate.

Nuclear Power Plant Emergency

If an accident occurs at a nuclear power plant near you, stay calm and listen to your local television or radio stations for updates and instructions from your state and local of officials.

The NRC and the nuclear power industry define emergencies according to four levels of increasing significance: Unusual Event, Alert, Site Area Emergency, and General Emergency. See Emergency Classification for the definitions of these four levels.

Only in a General Emergency is radiation expected to leave the plant site in amounts that might threaten public safety; however, local authorities might initiate protective actions at an earlier stage.

If you live within 10 miles of the plant, you are within the "Emergency Planning Zone." You may receive one or more alerts to warn you of the situation once protection actions are initiated:

- Sirens

- A tone alert on your radio

- A route alert from emergency responders (the "Paul Revere" method)

- A robocall from a public safety warning system (e.g., REVERSE 911®)

If you hear such a warning, tune your radio or television to the Emergency Alert System station for your area. The station is identified in the emergency preparedness information you receive annually from the utility.

Follow the instructions you receive from this station.

These instructions may include directions for evacuating or for remaining where you are (called sheltering in place) to reduce any possible exposure to radiation. Remember:

- Instructions will come from your state and local officials, not the NRC

- Evacuation may not be the best course of action, depending on the conditions of the emergency and other factors such as wind direction.

In serious situations, federal response information will be available on the NRC website, the NRC blog, and the NRC Twitter feed, as well as the Federal Emergency Management Agency (FEMA)'s ready.gov.

Remember: If you are hurt or sick and require immediate help, call 911.

If you live near the plant but outside the 10 mile Emergency Planning Zone, you most likely will not be asked or required to take protective action. Stay tuned to your local media for up to the minute information and advice from your state and local officials.

Dirty Bomb

A dirty bomb, or "radiological dispersal device," combines a conventional explosive such as dynamite with a radioactive material. If one explodes near you, take the following steps:

- Move at least several blocks from the explosion and get indoors. This will reduce exposure to any airborne radioactive dust.
- Check local radio or television stations for advisories from emergency response and health authorities.
- Remove your clothes and place them in a sealed plastic bag. Your contaminated clothing can be tested for radiation contamination.
- Take a shower to wash off dust and dirt. This will reduce total radiation exposure, if the explosive device contained radioactive material.
- If radioactive material was released, local news broadcasts will advise you where to report for radiation monitoring and tests to determine whether you were exposed and what steps to take to protect your health.

Source: Nuclear Regulatory Commission. "What Do I Do in a Nuclear Emergency?" January 2013. http://www.nrc .gov/about-nrc/emerg-preparedness/in-radiological-emerg .html

Debating the Future of Nuclear Power

The Nuclear Renaissance (World Nuclear Association)

The World Nuclear Association has continued to make a case for the expansion of nuclear power both as a means of meeting growing energy demand and as a way to reduce use of fuels such as coal and oil that create greenhouse gases.

Increasing energy demand, plus concerns over climate change and dependence on overseas supplies of fossil fuels are coinciding to make the case for increasing use of nuclear power.

- China is embarking upon a huge increase in nuclear capacity to 70-80 GWe by 2020; India's target is to add 20 to 30 new reactors by 2020.
- Communities in Finland and Sweden have accepted the local construction of permanent disposal sites for nuclear waste.
- International cooperation and commerce in the field of nuclear science and technology is growing.
- A WNA projection shows at least 1100 GWe of nuclear capacity by 2060, and possibly up to 3500 GWe, compared with 373 GWe today.
- Most of this increase (over 80%) will be in countries which already use nuclear power.

Since about 2001 there has been much talk about an imminent nuclear revival or "renaissance" which implies that the nuclear industry has been dormant or in decline for some time. Whereas this may generally be the case for the Western world, nuclear capacity has been expanding in Eastern Europe and

Asia. Globally, the share of nuclear in world electricity has showed slight decline from about 17% to 13.5% since the mid 1980s, though output from nuclear reactors actually increased to match the growth in global electricity consumption.

Today nuclear energy is back on the policy agendas of many countries, with projections for new build similar to or exceeding those of the early years of nuclear power. This signals a revival in support for nuclear power in the West that was diminished by the accidents at Three Mile Island and Chernobyl and also by nuclear power plant construction cost overruns in the 1970s and 1980s, coupled with years of cheap natural gas.

The March 2011 Fukushima accident has set back public perception of nuclear safety, despite there being no deaths or serious radiation exposure from it (while the direct death toll from the tsunami which caused it is some 25,000). Also the advent of shale gas has adversely changed the economics of nuclear power in places such as North America.

Drivers for the Nuclear Renaissance

The first generation of nuclear plants were justified by the need to alleviate urban smog caused by coal-fired power plants. Nuclear was also seen as an economic source of base-load electricity which reduced dependence on overseas imports of fossil fuels. Today's drivers for nuclear build have evolved:

Increasing energy demand

Global population growth in combination with industrial development will lead to a doubling of electricity consumption from 2007 level by 2030. Besides this incremental growth, there will be a need to replace a lot of old generating stock in the USA and the EU over the same period. An increasing shortage of fresh water calls for energy-intensive desalination plants, electric vehicles will increase overnight (base-load) demand, and in the longer term hydrogen production for transport purposes will need large amounts of electricity and/or high temperature heat.

Security of Supply

A re-emerging topic on many political agendas is security of supply, as countries realize how vulnerable they are to interrupted deliveries of oil and gas. The abundance of naturally occurring uranium and the large energy yield from each tonne of it makes nuclear power attractive from an energy security standpoint. A year or two's supply of nuclear fuel is easy to store and relatively inexpensive.

Climate change

Increased awareness of the dangers and effects of global warming and climate change has led decision makers, media and the public to agree that the use of fossil fuels must be reduced and replaced by low-emission sources of energy. Popular sentiment focuses on renewables, but nuclear power is the only readily-available large-scale alternative to fossil fuels for production of continuous, reliable supply of electricity (ie meeting base-load demand).

Economics

Increasing fossil fuel prices have greatly improved the economics of nuclear power for electricity now. Several studies show that nuclear energy is the most cost-effective of the available base-load technologies, at least when natural gas prices are high. In addition, as carbon emission reductions are encouraged through various forms of government incentives and emission trading schemes, the economic benefits of nuclear power will increase further.

Insurance against future price exposure

A longer-term advantage of uranium over fossil fuels is the low impact that increased fuel prices will have on the final electricity production costs, since a large proportion of those costs is in the capital cost of the plant. This insensitivity to fuel price fluctuations offers a way to stabilize power prices in deregulated markets.

As the nuclear industry is moving away from small national programmes towards global cooperative schemes, serial production of new plants will drive construction costs down, as is already being shown in China, and further increase the competitiveness of nuclear energy.

An enabling factor is the increasing ability of nuclear reactors to load-follow, adjusting their output according to demand, so that they are less restricted to steady base-load role. However in the short term this is only relevant where nuclear power supplies more than about 60% of the power.

In practice, is a rapid expansion of nuclear power capacity possible?

Most reactors today are built in under five years (first concrete to first power), with four years being state of the art and three years being the aim with modular prefabrication. Several years are required for preliminary approvals before construction.

It is noteworthy that in the 1980s, 218 power reactors started up, an average of one every 17 days. These included 47 in USA, 42 in France and 18 in Japan. The average power was 923.5 MWe. So it is not hard to imagine a similar number being commissioned in a decade after about 2015. But with China and India getting up to speed with nuclear energy and a world energy demand double the 1980 level in 2015, a realistic estimate of what is possible might be the equivalent of one 1000 MWe unit worldwide every 5 days.

A relevant historical benchmark is that from 1941 to 1945, 18 US shipyards built over 2700 Liberty Ships. These were standardised 10,800 dwt cargo ships of a very basic British design but they became symbolic of US industrial wartime productivity and were vital to the war effort. Average construction time was 42 days in the shipyard, often using prefabricated modules.* In 1943, three were being completed every day. They were 135 metres long and could carry 9100 tonnes of cargo, so comparable in scale if not sophistication to nuclear reactors.

*As a publicity stunt, and using a lot of prefabrication, in 1942 the Robert G. Peary was launched in under five days and ready for sea three days later.

Public acceptance

During the early years of nuclear power, there was a greater tendency amongst the public to respect the decisions of authorities licensing the plants, but this changed for a variety of reasons. No revival of nuclear power is possible without the acceptance of communities living next to facilities and the public at large as well as the politicians they elect.

The 1986 Chernobyl disaster marked the nadir of public support for nuclear power. However, this tragedy underscored the reason for high standards of design and construction required in the West. It could never have been licensed outside the Soviet Union, incompetent plant operators exacerbated the problem, and partly through Cold War isolation, there was no real safety culture. The global cooperation in sharing operating experience and best practices in safety culture as a result of the accident has been of benefit worldwide. The nuclear industry's safety record over the next 25 years helped restore public faith in nuclear power, though the multiple-reactor Fukushima accident in 2011 brought his run of unrivalled safety to an end. Over this period, operating experience tripled, from about 4000 reactor-years to more than 14,500 reactor years (plus a similar total in the nuclear navies). But an objective look at the safety record, despite Fukushima, still shows it very favourably compared with any alternative.

Another factor in public reassurance is the much smaller than anticipated public health effects of the Chernobyl accident. At the time some scientists predicted that tens of thousands would die as a result of the dispersal of radioactive material. In fact, according to the UN's Chernobyl Forum report, as of mid 2005, fewer than 60 deaths had been directly attributed to radiation from the disaster, and further deaths from cancer are uncertain. The human toll from Fukushima is no deaths from the nuclear accident and no serious radiation effects, but the inconvenience of evacuation has been massive, just as it was with Chernobyl.

One of the criticisms often levelled against nuclear power is the alleged lack of strategy and provision for its long-lived wastes. It is argued that local communities would never be prepared to host a repository for such waste. However, experience has shown in Sweden and Finland, that with proper consultation and compensation, mostly in the form of long-term job prospects, communities are quite prepared to host repositories. Indeed in Sweden, two communities were competing to be selected as the site of the final repository. In fact, radioactive wastes, including those from the nuclear industry, are handled and managed responsibly in all countries, and there has never been any harm or hazard to anyone on account of those from nuclear power.

New nuclear power capacity

With 60 reactors being built around the world today, another 150 or more planned to come online during the next 10 years, and over two hundred further back in the pipeline, the global nuclear industry is clearly going forward strongly. Responses to the Fukushima accident, notably in Europe, do not change this overall picture. Countries with established programmes are seeking to replace old reactors as well as expand capacity, and an additional 25 countries are either considering or have already decided to make nuclear energy part of their power generation capacity. However, most (over 80%) of the expansion in this century is likely to be in countries already using nuclear power.

A World Nuclear Association exercise "Nuclear Century Outlook" projects possible expansion in world nuclear generating capacity. From a base of 377 GWe today it projects at least 1130 GWe by 2060 and up to 3500 GWe by then. The upper projection for 2100 is 11,000 GWe.

Source: World Nuclear Association. "The Nuclear Renaissance," August 2011. http://www.world-nuclear .org/info/Current-and-Future-Generation/The-Nuclear -Renaissance/#.UWY5Z5DxakB. Used by permission.

Nuclear Power and Climate: Why Nukes Can't Save the Planet

The Nuclear Information and Resource Service, which primarily serves antinuclear activists, offers a cogent summary of the arguments against expanding nuclear power as a means to address climate change.

Too Many Reactors; Not Enough Carbon Reductions

Major studies (from MIT, Commission on Energy Policy, and International Atomic Energy Agency, for example) agree that about 1,500-2,000 large new atomic reactors would have to be built for nuclear power to make any meaningful dent in greenhouse emissions. Operation of that many new reactors (currently about 440 exist worldwide) would cause known uranium reserves to run out in just a few decades and force mining of lower-grade uranium, which itself would lead to higher greenhouse emissions. If all of these reactors were used to replace coal plants, carbon emissions would drop by about 20% worldwide. If used entirely as new capacity, in the place of sustainable technologies like wind power, solar power, energy efficiency, etc., carbon emissions actually would increase.

Too Much Money

Construction of 1,500 new reactors would cost trillions of dollars. U.S. reactors going online in the 1980s and 90s averaged about $4 billion apiece. Cost estimates for a new generation of reactors range from $5 billion to $9 billion or more each. Use of resources of this magnitude would make it impossible to also implement genuinely effective means of addressing global warming. Energy efficiency improvements, for example, are seven times more effective at reducing greenhouse gases, per dollar spent, than nuclear power.

Too Much Time

Construction of 1,500 new reactors would mean opening a new reactor about once every two weeks, beginning today, for the next 60 years—an impossible schedule. The world's nuclear reactor manufacturers currently are capable of building about half that amount. Since reactors take 6-10 years to build (some U.S. reactors that began operation in the 1990s took more than 20 years), we are already that long behind schedule and will fall farther behind. Addressing the climate crisis cannot wait for nuclear power.

Too Much Waste

Operation of 1,500 or more new reactors would create the need for a new Yucca Mountain-sized radioactive waste dump somewhere in the world every 3-4 years. Yucca Mountain has been under study for nearly 20 years, has been vigorously opposed by the State of Nevada for just as long, and remains at least a decade from completion.

The odds of identifying numerous new scientifically-defensible and publicly-acceptable waste dumps are slim.

International efforts to site radioactive waste facilities are similarly behind schedule and face substantial public opposition. For this reason, the U.S. and other countries are attempting to increase reprocessing of nuclear fuel as a waste management tool—a dangerous and failed technology that increases worldwide nuclear proliferation risks.

Too Little Safety

Odds of a major nuclear accident are on the order of 1 in 10,000 reactor-years. Operation of some 2,000 reactors (1,500 new plus 440 existing) could result in a Chernobyl-scale nuclear accident as frequently as every five years—a price the world is not likely to be willing to pay. Reactors of similar designs likely would close following a major accident, making nuclear power a risky proposition as a climate solution. And more reactors means more potential terrorist targets.

Too Much Plutonium

Operation of 1,500 or more new reactors would require a dozen or more new uranium enrichment plants, and would result in the production of thousands of tons of plutonium (each reactor produces about 500 pounds of plutonium per year), posing untenable nuclear proliferation threats.

Nukes Emit Carbon Too!

While atomic reactors themselves are not major emitters of greenhouse gases, the nuclear fuel chain produces significant greenhouse emissions. Besides reactor operation, the chain includes uranium mining, milling, processing, enrichment, fuel fabrication, and long-term radioactive waste storage, all of which are essential components of nuclear power. At each of these steps, construction and operation of nuclear facilities results in greenhouse gas emissions. The uranium enrichment plant at Paducah, Kentucky, for example, is the largest U.S. emitter of ozone-destroying ChloroFluoroCarbons (CFCs)— banned by the Montreal Protocol (the Paducah plant was grandfathered by this treaty).

Taken together, the fuel chain greenhouse emissions approach those of natural gas—and are far higher than emissions from renewable energy sources, not to mention emissions-free energy efficiency technologies.

Not Suited for Warming Climates

Unlike solar power, nuclear power does not work well in warming climates. The summer of 2004's heat wave across Europe not only killed thousands of people, but because of dwindling river levels caused many reactors to reduce power levels and even shut down entirely. Reactors require vast quantities of water to keep the core cool; changes in water levels, and even water temperatures, can greatly affect reactor operations. Reactors in the U.S. have similarly been forced to close during heat waves.

Can't Take Us to the Mall

Nuclear power, which can only produce electricity, does not address emissions from automobiles and other components of the transportation sector—one of the largest source of carbon emissions.

What We Can Do: A Carbon-Free, Nuclear-Free Future

Major investment in energy supply will be needed to meet growing energy demand and address the climate crisis at the same time—perhaps even as much as building 1,500 new reactors would cost.

But investing the money differently gives us much more bang for the buck: instead of a 20% reduction in carbon emissions, we can get a 100% reduction—and that's a goal worth working for.

By 2050, the world will need about 25-30 Terrawatts of energy, or the equivalent of 25-30,000 nuclear reactors. Clearly it is not possible or affordable to build that many reactors. But it is possible to build that much capacity through energy efficiency improvements, and sustainable energy sources including wind, biomass, geothermal, and especially solar power—if we start making the necessary investments now. It won't be cheap or easy, but the payoff is huge: safe, clean energy that helps alleviate rather than contribute to the climate crisis. Our choice is stark: we can choose nuclear power, or we can address global warming. We can't do both. Fortunately, the choice is an easy one.

Source: Nuclear Information and Resource Service. "Nuclear Power and Climate: Why Nukes Can't Save the Planet." http://www.nirs.org/factsheets/nukesclimatefact 208.pdf. Used by permission.

One might say that over the past four decades or so, there have been several "waves" of publications relating to nuclear power. In the 1970s and 1980s the near disaster at Three Mile Island and the undoubted disaster that was Chernobyl led to a vigorous debate over the risks and future of nuclear power. During the early to mid-1980s, this debate would overlap concern with heightened Cold War tensions, growing nuclear arsenals, and the likely impacts of nuclear war.

With the collapse of the Soviet Union, these concerns faded somewhat from public consciousness, while among experts and policy makers questions of nuclear proliferation and the disposition of the former Soviet nuclear arsenal came to the fore. The terrorist attacks of September 11, 2001, would heighten the already considerable concern about terrorists obtaining or possibly constructing nuclear or radiological weapons.

Meanwhile, some industry and outside advocates began to speak of a "nuclear renaissance" brought about by growing concerns related to global climate change and the need to substantially reduce the use of fossil fuels and their carbon emissions. However, in 2011 the Fukushima earthquake, tsunami, and resulting nuclear reactor meltdowns would again raise the

---

The massive 1986 Chernobyl nuclear disaster resulted in the declaration of an "exclusion zone" of about 1000 square miles (2600 square km.) Due to continuing radioactivity, only specialized workers are generally allowed to enter. (Shutterstock)

question of whether nuclear power could ever be safe in a future likely to be marked by more extreme climatic events.

The print and online materials described in this chapter are intended to provide a useful (if not exhaustive) selection of sources addressing all of these concerns from a variety of viewpoints. The materials have been organized into the following topics, with separate sections in each topic for print and online materials:

- Overview and Reference
- Historical Developments
- Economic, Regulatory, and Legal Aspects
- Environmental and Health Effects
- Risks, Disasters, and Safety in Nuclear Plants
- Nuclear Waste and the Fuel Cycle
- Proliferation and Terrorism
- New Designs and Technologies
- Climate Change and the Future of Nuclear Power

Overview and Reference

Works in this section provide general overview, background, or reference involving more than one topic in nuclear energy. For works involving particular historical periods or developments, see the next section.

Print Materials

Angelo, Joseph A. *Nuclear Technology*. Westport, CT: Greenwood, 2004.

> A comprehensive reference to all aspects of nuclear science and technology, including biographical profiles, chronology, glossary, and guide to resources.

Atkins, Steven. *Historical Encyclopedia of Atomic Energy.* Westport, CT: Greenwood, 2000.

> An extensive reference to the people, concepts, discoveries, and events that created the modern world of nuclear energy. Includes extensive coverage of key events, including nuclear accidents as well as the development of regulations and treaties.

Bodansky, David. *Nuclear Energy: Principles, Practices, and Prospects.* 2nd ed. New York: Springer-Verlag, 2010.

> Definitely textbook-like rather than a popular account, but a clear, well-organized, and comprehensive account of nuclear science and technology, including current developments and issues. Readers with some basic background in physics would find it easier going in places.

Diehl, Sarah J., and James Clay. *Nuclear Weapons and Nonproliferation: A Reference Handbook.* 2nd ed. Santa Barbara, CA: ABC-CLIO, 2007.

> Explores the historical and contemporary issues involved with nuclear deployment, disarmament, and proliferation around the world, with full coverage of "hot spots" such as Iran, Pakistan, and North Korea. Includes a selection of source documents, detailed chronology, biographical sketches, and bibliography.

Ezhela, V. V. (ed.). *Particle Physics: One Hundred Years of Discoveries: An Annotated Chronological Bibliography.* Woodbury, NY: American Institute of Physics, 1996.

> Includes not only bibliographical information, but overall context for the key discoveries in nuclear physics that led to nuclear power (and much else). Also includes abstracts and a chronological list of discoveries.

Ferguson, Charles D. *Nuclear Energy: What Everyone Needs to Know.* New York: Oxford University Press, 2011.

A top physicist and nuclear engineer provides an accessible, even-handed overview of the science, technology, and issues surrounding nuclear energy via a question-and-answer format.

Irvine, Maxwell. *Nuclear Power: A Very Short Introduction*. New York: Oxford University Press, 2011.

A good overview of the underlying physics and development of the nuclear power industry, along with concerns related to safety, waste disposal, and economic viability.

Makhijani, Arjun, and Scott Saleska. *The Nuclear Power Deception: U.S. Nuclear Mythology from Electricity "Too Cheap to Meter" to "Inherently Safe" Reactors*. New York: Apex, 1999.

The first part of the book surveys the early developments of nuclear power and the promises made by its proponents. The authors go on to look at what has been learned from accidents and offer a critique of allegedly safer new reactor designs. Finally, nuclear power is viewed in the light of new energy challenges and the Kyoto Protocol on global warming.

Pilat, Joseph F. (ed.). *Atoms for Peace: A Future after Fifty Years?* Baltimore: Johns Hopkins Press, 2007.

A collection of papers published on the fiftieth anniversary of Eisenhower's famous "Atoms for Peace" speech. A variety of topics are covered, including the legacy of the Eisenhower program, the promise and future of nuclear energy, and ongoing concerns with proliferation and terrorism.

Tuniz, Claudio. *Radioactivity: A Very Short Introduction*. New York: Oxford University Press, 2012.

Any intelligent discussion of nuclear issues requires a basic understanding of radioactivity. This introduction explains this natural phenomena and discusses its many uses, including scientific research, medical treatment, industry, and power production.

Web Resources

ABCs of Nuclear Science. Lawrence Berkeley National Laboratory.

> http://www.lbl.gov/abc/
> An introduction to nuclear science geared to students and teachers. Includes a variety of experiments and activities that can be used in school laboratories, using safe radioactive sources.

ALSOS Digital Library for Nuclear Issues.

> http://alsos.wlu.edu/
> An extensive collection of references to resources (books, articles, multimedia, and websites) related to nuclear issues. A detailed search form is provided. Topics include extensive coverage of the Manhattan Project and its aftermath, nuclear weapons and the Cold War, nuclear power, and contemporary issues of nuclear proliferation and terrorism.

The Language of the Nucleus: a Glossary of 100,000+ Nuclear Science, Engineering, Medicine, and National Security Terms.

> wwww.nuclearglossary.com
> Created and Maintained by Craig Stone of Scientific Digital Visions, this is probably the largest imaginable glossary of terms relevant to all nuclear topics. The glossary is built using an XML database and links base or root terms to synonyms or variants.

"Nuclear Education, Research and Careers." American Nuclear Society.

> http://www.aboutnuclear.org/erc/
> A guide to a variety of careers accessible by earning a degree in nuclear engineering or health physics. They include nuclear power and propulsion, numerous medical applications, and specialties such as nuclear security and nonproliferation.

Nuclear Energy Guide. Council on Foreign Relations.
> http://www.cfr.org/interactives/IG_Nuclear/index.
> html#/overview/
> A nicely organized and richly illustrated interactive guide
> to nuclear energy, including an overview, timeline, cur-
> rent developments and analysis (with many video inter-
> views), future prospects, and links to resources.

Nuclear Energy. North Carolina State University.
> https://sites.google.com/a/ncsu.edu/nuclear-energy/home
> An extensive and valuable resource site built by a group of
> engineering graduate students and others. Provides back-
> ground and links to many aspects of nuclear power,
> including history, technology, economics, environment,
> public perception, and domestic and international
> politics.

Nuclear Energy Resources for Schools. Argonne National
Laboratory.
> http://students.ne.anl.gov/schools/
> An excellent collection of online links to nuclear energy
> topics ranging from the basic (e.g., What is radioactivity?)
> to how nuclear power plants work waste management.
> Designed primarily for junior high and high school
> students.

Nuclear Power Information Tracker. Union of Concerned
Scientists.
> http://www.ucsusa.org/nuclear_power/reactor-map/
> embedded-flash-map.html
> An interactive map showing nuclear plants (active or
> decommissioned) in the United States. Clicking on a loca-
> tion gives basic information, including type, current sta-
> tus, and reported safety issues.

"Nuclear Science and Technology and How It Influences Your
Life." American Nuclear Society.

http://www.aboutnuclear.org/home.cgi
A guide to aspects and applications. Includes basic information and a glossary. As one might expect, it takes a pro-nuclear perspective.

Historical Developments

Works in this section deal with the full sweep of the historical development of nuclear energy or with particular periods ore events such as early nuclear discoveries, the Manhattan Project, "Atoms for Peace" and the Cold War

Print Materials

Beaver, William. *Nuclear Power Goes On-Line: A History of Shippingport*, New York: Greenwood Press, 1990.

A detailed account of the construction of Shippingport, the first commercial nuclear power plant to go online. Through the lens of this case study, the author examines how the nuclear industry and government struggled with technical and regulatory issues.

Bernstein, Jeremy. *Plutonium: A History of the World's Most Dangerous Element*. Washington, DC: Joseph Henry, 2010.

Traces the element from its discovery as a theoretical novelty in 1941 to its key role as the fuel for nuclear weapons and atomic power. Includes lively descriptions of the work of early atomic physicists, followed by a more controversial assessment of the risks and threats arising from the world's plutonium stockpiles.

Bickel, Lennard. *The Deadly Element: The Story of Uranium.* New York: Stein and Day, 1979.

A fascinating account of uranium from its discovery in 1763 to the discoveries of early nuclear physicists to the transmutation of elements and the creation of plutonium in the Manhattan Project by bombarding uranium.

Burns, Richard Dean, and Joseph M. Siracusa. *A Global History of the Nuclear Arms Race*. Santa Barbara, CA: Praeger, 2013.

> Many of the issues surrounding "peaceful" nuclear power are inextricably bound up with the development, deployment, and proposed limitations of nuclear weapons. In this book two leading experts chart the development of nuclear strategy and issues through the end of the Cold War and beyond.

Hacker, Barton C. *The Dragon's Tail: Radiation Safety in the Manhattan Project, 1942–1946*. Berkeley: University of California Press, 1987.

> This account details the slow development of awareness of the dangers of radiation and of safety measures during the development and first tests of the atomic bomb.

Kelly, Cynthia C. (ed.). *The Manhattan Project: The Birth of the Atomic Bomb in the Words of Its Creators, Eyewitnesses, and Historians*. New York: Black Dog & Leventhal, 2007.

> Compiled by the Atomic Heritage Foundation, a rich trove of original sources by eyewitnesses who played key roles in the development of nuclear weapons and power. Includes reflections and reactions following the dropping of the atomic bombs in Japan.

Kiernan, Denise. *The Girls of Atomic City: The Untold Story of the Women Who Helped Win World War II*. New York: Touchstone, 2013.

> The remarkable story of the many women who worked at Oak Ridge, the massive facility that processed uranium for the first atomic bombs, in an atmosphere of secrecy and little-understood dangers.

Lanouette, William, and Bela Silard. *Genius in the Shadows: A Biography of Leo Szilard, The Man behind the Bomb*. New York: Scribner's, 1992.

Biography of Leo Szilard, the Hungarian American physicist who first conceived of the nuclear chain reaction and then wrote the letter that Einstein sent to Franklin Roosevelt in 1939, warning of the possibility of extremely destructive new bombs and the possibility that Germany was already working on such a weapon. After looking at his participation in (and increasing disillusionment with) the Manhattan Project, the authors also describe Szilard's later career in which he worked for nuclear disarmament and shifted his scientific interests from physics to biology.

Mahaffey, James. *Atomic Awakening: A New Look at the History and Future of Nuclear Power.* New York: Pegasus, 2010.

A vivid, fascinating account of the development of nuclear power. The author says his intention is "not to sell nuclear power," but to provide a balanced history (including accidents and dangers) that he believes will set the debate on a new course that might recognize that nuclear power is needed for future energy needs and can be operated safely.

Marcus, Gail H. *Nuclear Firsts: Milestones on the Road to Nuclear Power.* La Grange Park, IL: American Nuclear Society, 2010.

An interesting approach to nuclear history that focuses on the "firsts," or key developments, that would shape the development of the nuclear industry. These include discoveries in nuclear physics, the first fission and reactor experiments, the first commercial nuclear power plants, and other applications of nuclear energy.

Preston, Diana. *Before the Fallout: From Marie Curie to Hiroshima.* New York: Walker, 2005.

This is a lively account by a noted historian of the different strands of science and discovery in nuclear physics that would eventually produce fission and the atomic bomb. Biographies of pioneers such as the Curies, Rutherford, and Bohr are woven into the narrative.

U.S. Department of Energy. *Hanford Site Historical District: History of the Plutonium Production Facilities, 1943–1990.* Columbus, OH: Battelle, 2003.

> A detailed narrative of the siting, construction, and operation of the Hanford site, a major source of plutonium for nuclear weapons from the Manhattan Project until the end of the Cold War. The account is well illustrated and rich in personal material, and it includes an extensive bibliography.

Weart, Spencer R. *The Rise of Nuclear Fear.* Cambridge, MA: Harvard University Press, 2012.

> This book provides a neglected perspective on nuclear issues. It offers a sort of cultural history of our relationship to the exploding atom, radioactivity, nuclear weapons, and nuclear power over the past half century or so. The depiction of these matters in images, symbols, and popular culture brings vividness to the issues discussed.

Wellock, Thomas. *Critical Masses: Opposition to Nuclear Power in California, 1958–1978.* Madison: University of Wisconsin Press, 1998.

> Interviews, oral history, and original documents paint a picture of the rise of the antinuclear movement in California. The author, a nuclear engineer turned academic historian, suggests that the political experience of activists and the issues brought to public consciousness played an important role in the development of a broader environmental coalition.

Wills, John. *Conservation Fallout: Nuclear Protest at Diablo Canyon.* Reno: University of Nevada Press, 2012.

> The controversy over nuclear power at Diablo Canyon brought together many key issues, ranging from the risks of building in an earthquake-prone region to the potential effects on some of California's most fertile agricultural

land. According to the author, the upshot of this social and cultural history was "fallout" in the form of new impetus for the environmental and conservation movements.

Web Resources

The Acheson-Lilienthal Report on the International Control of Atomic Energy.

> http://www.learnworld.com/ZNW/LWText.Acheson-Lilienthal.html
> Full text of the seminal report of a committee on atomic energy chaired by Deach Acheson and David Lilienthal in 1946. The report suggests that because the secrets of atomic energy could not be reasonably assumed to remain a monopoly of the United States and its allies, a mechanism under the recently formed United Nations should be given control over nuclear matters, including the stockpile of fissionable materials.

Atomic Archive. AJ Software and Multimedia.

> http://www.atomicarchive.com/
> A companion to a CD product but rather complete in itself, this site offers overviews of historic developments, chronology, media, and resources for further research.

"Atoms for Peace" address by President Dwight Eisenhower.

> http://www.nuclearfiles.org/menu/key-issues/nuclear-energy/history/atoms-for-peace-address-eisenhower_1953-12-08.htm
> The famous address by Eisenhower to the UN General Assembly on December 8, 1953. It proposed international controls on nuclear materials and the promotion of a civilian nuclear power industry.

Chater, James. "A History of Nuclear Power." Focus on Nuclear, 2005.

http://www.focus-nuclear.com/pdf/TP_History_nuclear.pdf
An illustrated account of the development of nuclear power, suggesting that it has a "moderately bright" future.

The First Reactor. U.S. Department of Energy, 1982.
http://local.ans.org/mi/Teacher_CD/Historical%20Info/
DOE-NE-0046.pdf
Describes the building of "the Pile"—the first nuclear reactor. Includes theory and designs, organization of the Manhattan Project, and personal accounts by Enrico and Laura Fermi of life during the project.

"Moments of Discovery: The Discovery of Fission." American Institute of Physics.
http://aip.org/history/mod/fission/fission1/01.html
Highlights the ideas and personalities involved in early nuclear research leading to the discovery of fission. Includes audio files in which many of these pioneers recount their work.

Outline History of Nuclear Energy. World Nuclear Association.
http://www.world-nuclear.org/info/Current-and-Future-
Generation/Outline-History-of-Nuclear-Energy/
Good summary of the development of nuclear power, including the development of reactors and power plants. The account admits that nuclear power "suffered some decline and stagnation" from the late 1970s to about 2002 but suggests that a "nuclear renaissance" may be in the offing.

Savannah River Site at Fifty. U.S. Department of Energy.
http://www.srs.gov/general/about/50anniv/
CONTENTS.pdf
Online version of a 2003 book that recounts the development and operation of the U.S. Department of Energy Savannah River Site in South Carolina. From 1950 to 2000 this facility was the scene for important developments

in reactor technology, production of fuel for nuclear weapons, waste management, and procedures for decommissioning nuclear plants. Today work is focused on waste processing, including extraction and reprocessing.

Economic, Regulatory, and Legal Aspects

Items in this section deal with the often interconnected questions of the economic viability of nuclear power, the effects of regulation on the cost of nuclear power and its attractiveness as an investment, and issues relating to tax and insurance subsidies received by the nuclear industry.

Works primarily dealing with the future desirability of nuclear power will be found in the final section, "Climate Change and the Future of Nuclear Power."

Print Materials

Aron, Joan. *Licensed to Kill? The Nuclear Regulatory Commission and the Shoreham Power Plant.* Pittsburg: University of Pittsburg Press, 1997.

> A history of the problem-plagued Shoreham nuclear power plant in Long Island, New York. It attempts to untangle the complex political and legal struggle between the NRC, the state government, and public activist groups. After the developers had spent $6 billion, the plant was shut down without ever having produced any electricity for consumers.

Cohen, Martin, and Andrew McKillop. *The Doomsday Machine: The High Price of Nuclear Energy, the World's Most Dangerous Fuel.* New York: Palgrave Macmillan, 2012.

> An environmentalist/social scientist and an energy economist team up to argue that the true costs of nuclear power have largely been hidden from us—including the impact of uranium mining and the cleanup costs following a major accident or meltdown.

Cooke, Stephanie. *In Mortal Hands: A Cautionary History of the Nuclear Age*. New York: Bloomsbury USA, 2009.

> An interesting anecdote-filled narrative that raises issues thoughtfully, taking a generally skeptical attitude toward the ability of government to regulate and restrain the growth of a "nuclear industrial complex."

Ford, Daniel F. *The Cult of the Atom: The Secret Papers of the Atomic Energy Commission*. Revised ed. New York: Simon & Schuster, 1984.

> The author used the Freedom of Information Act to unearth documentation for his history of the AEC as it faced many nuclear "incidents" (generally unknown to the public) and, he argues, usually sided with the industry against the public interest and safety.

Hewlett, Richard G., and Oscar E. Anderson. *The New World: A History of the United States Atomic Energy Commission. Vol. 1. 1939–1946*. University Park: Pennsylvania State University Press, 1962.

> The first part of the official and definitive history of America's nuclear regulatory agency. This is actually the prehistory focusing on the Manhattan Project and the technical developments and issues that would need to eventually be addressed by regulators.

Hewlett, Richard G., and Francis Duncan. *Atomic Shield: A History of the United States Atomic Energy Commission. Vol. 1I, 1947–1952*. University Park: Pennsylvania State University Press, 1969.

> The history continues as the Atomic Energy Commission was born while many of the technical problems of nuclear power were still being worked out and an apparatus was needed to turn "atoms for peace" into a viable and manageable nuclear industry.

Hewlett, Richard G., and Jack M. Holl. *Atoms for Peace and War, 1953-1961: Eisenhower and the Atomic Energy Commission.* Berkeley: University of California Press, 1989.

> The history concludes with the beginnings of the establishment of nuclear power, the struggle to contain the Cold War, and other aspects of implementing the ambitious nuclear program of the Eisenhower administration.

Martin, Richard. *Super Fuel: Thorium, the Green Energy Source for the Future.* New York: Palgrave Macmillan, 2012.

> Thorium has been a long overlooked alternative to uranium and plutonium as a fuel for nuclear reactors. The author argues that thorium's abundance could put the nuclear industry on a much more economical and stable footing. Further, it is argued that thorium would make safer reactors possible because using it does not result in the production of fission products that are dangerous and potentially subject to proliferation or diversion by terrorists.

Mez, Lutz, Mycle Schneider, and Steve Thomas (eds.). *International Perspectives on Energy Policy and the Role of Nuclear Power.* Essex, United Kingdom: Multi-Science, 2009.

> A collection of case studies of 30 countries with nuclear power programs, broken into three categories: those undergoing a "nuclear renaissance" (expansion), those attempting to phase out nuclear power, and newly industrializing nations. Numerous graphs and charts enable a detailed picture of the relationship between nuclear power and the energy economy of each country.

Pope, Daniel. *Nuclear Implosions: The Rise and Fall of the Washington Public Power Supply System.* New York: Cambridge University Press, 2008.

> In the 1970s Washington State tried to build a huge nuclear power complex with five reactors. The challenge

proved overwhelming to the unfortunately nicknamed "Whoops" (Washington Public Power Supply System) and after becoming mired in construction and labor problems, bureaucracy, and political and environmental challenges, the project collapsed and led to defaults on $2.25 billion in municipal bonds. The author cites a variety of opinions from economists, political scientists, and others to offer explanations for the debacle and its implications for other nuclear power developments.

Sovacool, Benjamin K. *Contesting the Future of Nuclear Power: A Critical Global Assessment of Atomic Energy.* Hackensack, NJ: World Scientific, 2011.

A thorough critique that argues that building new nuclear plants is a lengthy, uneconomical process that must be subsidized by governments, relies on a diminishing supply of fuel, and is prone to unacceptable risks and environmental effects.

Sovacool, Benjamin K. *The National Politics of Nuclear Power: Economics, Security, and Governance.* New York: Routledge, 2012.

Provides a framework for understanding the factors involved in the development of nuclear power in nations with quite different domestic political situations. The author suggests that nations with centralized institutional power and a technocratic ideology are more likely to create a large nuclear power industry.

Suppes, Galen J. and Truman Storvick (eds.). *Sustainable Nuclear Power.* Boston: Elsevier Academic, 2006.

The contributors to this collection argue that nuclear power can be made effectively sustainable by improving the nuclear fuel cycle, greatly reducing the amount of uranium needed, and generating new fuel. Safety issues are also addressed.

Thomson, David B. *A Guide to the Nuclear Arms Control Treaties.* Los Alamos, NM: Los Alamos Historical Society, 2001.

> Explains the key international agreements on the control of nuclear weapons and materials, including the Nuclear Non-Proliferation Treaty (NPT).

Vandenbosch, Robert, and Susanne E. Vandenbosch. *Nuclear Waste Stalemate: Political and Scientific Controversies.* Salt Lake City: University of Utah Press, 2007.

> A husband and wife team (a chemist and a political scientist) provide a detailed account of the political controversies over siting a nuclear waste repository that stalled the Yucca Mountain project and the scientific disputes over the safest way to contain waste and the risks posed by seismic and volcanic events.

Web Resources

Holt, Mark. "Nuclear Energy Policy." Congressional Research Service, 2012.

> http://www.fas.org/sgp/crs/misc/RL33558.pdf
> Summarizes current issues and the status of nuclear power policy, including industry incentives in the Energy Policy Act of 2005, the DOE's nuclear research and development programs, and ongoing attempts to provide for nuclear waste facilities in the wake of the cancellation of the Yucca Mountain project.

"Nuclear Power's Role in Generating Electricity." U.S. Congressional Budget Office, May 2008.

> http://cbo.gov/sites/default/files/cbofiles/ftpdocs/91xx/doc9133/05-02-nuclear.pdf
> An assessment of the competitiveness and future economic viability of nuclear power in the United States, including the effects of changes in the fuel markets and

possible regulation or taxing of carbon dioxide emissions. The lack of coverage of the impact of renewables such as solar or wind energy should be noted—these are not considered to be able to provide base load capacity for the foreseeable future.

"Renewable Energy and Electricity." World Nuclear Association. http://www.world-nuclear.org/info/Energy-and-Environment/Renewable-Energy-and-Electricity
A regularly updated position paper that surveys the current and likely future capacity for various forms of renewable energy (such as hydroelectric, wind, and solar).

Trubatch, Sheldon L. "How, Why, and When the U.S. Supreme Court Supports Nuclear Power." *Arizona Journal of Environmental Law & Policy*, October 4, 2012.
http://www.ajelp.com/wp-content/uploads/Trubatch_3ArizJEnvtlLPoly1.pdf
The author reviews the evolving jurisprudence of the U.S. Supreme Court on questions involving the regulation of nuclear power and related issues. At first the court deferred to the federal interest in regulating and promoting nuclear power as preempting state interests. Later, however, the court distinguished cases that it said did not directly relate to congressional interest and thus could be regulated by the states.

"A Short History of Nuclear Regulation, 1946–1999." Nuclear Regulatory Commission. http://www.nrc.gov/about-nrc/short-history.html#dawn
An overview of nuclear regulation, issues, and developments, with a particular focus on the Atomic Energy Commission and its successor, the Nuclear Regulatory Commission.

von Hippel, Frank (ed.). "The Uncertain Future of Nuclear Energy." A research report of the International Panel on Fissile Materials, September 2010.

http://fissilematerials.org/library/rr09.pdf
Evaluates data and trends affecting the economic and
political viability of nuclear power, including the economics of plant operation, decommissioning, and the fuel
cycle; the possible impact of new reactor designs; and
international attempts to control nuclear proliferation
and prevent terrorism. The report ends with recommendations to enhance nuclear safety and security.

Environmental and Health Effects

Materials in this section deal with the environmental impact
and health effects of radiation, such as from nuclear power
plants (normal operation or accidents), fallout from nuclear
tests, and natural radiation sources. For more on specific
nuclear accidents, see the later section titled "Risks, Disasters,
and Safety in Nuclear Plants."

Print Materials

Advisory Committee on Human Radiation Experiments. *Final
Report of the Advisory Committee on Human Radiation
Experiments*. New York: Oxford University Press, 1996.

> A fascinating and troubling as well as little-known part of
> America's atomic history involves many experiments that
> agencies such as the Department of Defense and the
> Atomic Energy Commission conducted involving the effects
> of radiation on human subjects. Many of these experiments
> would fall far short of today's ethical standards. These
> include administering radioactive materials to teenagers with
> developmental disabilities and hospital patients, as well as
> deliberately releasing radiation into the environment to
> study its dispersal and effects.

Allison, Wade. *Radiation and Reason: The Impact of Science on a
Culture of Fear*. York, UK: Wade Allison, 2009.

Offers an overview of radiation safety in relation to nuclear power, combined with advocacy for expansion of nuclear power as the most reliable solution to climate change. The author suggests that exaggerated fear of radiation and overestimation of risks have been part of the overall difficulty in educating the public about complex scientific issues.

Ansari, Armin. *Radiation Threats and Your Safety: A Guide to Preparation and Response for Professionals and Community*. Boca Raton, FL: Taylor & Francis / CRC, 2010.

A comprehensive guide to preparing for and responding to releases of radiation in the environment, both accidental and intentional. After giving a basic overview of radiation, the author looks at different types of radiation events and their possible health, environmental, and psychosocial effects. This is followed by coverage of various types of medical responses (including drugs to mitigate effects of radiation exposure) and advice for government officials, first responders, and concerned citizens.

Bergeron, Kenneth D. *Tritium on Ice: The Dangerous New Alliance of Nuclear Weapons and Nuclear Power*. Cambridge, MA: MIT Press, 2002.

The radioactive hydrogen isotope tritium is an essential component for hydrogen bombs. The author argues that the recent resumption of tritium production (by neutron absorption in nuclear power reactors) threatens increased nuclear proliferation. This is presented in a larger historical context of what the author sees to be inadequate safety measures and considerable environmental risks.

Brugge, Doug, Timothy Benally, and Esther Yazzie-Lewis (eds.). *The Navajo People and Uranium Mining*. Albuquerque: University of New Mexico Press, 2007.

A fascinating and little-told story of how the boom in uranium mining in the 1940s and 1950s brought economic opportunities to Navajo people but left a legacy of more than 1,000 abandoned uranium mines and mills and their contaminated surroundings. The book resulted from the Navajo Uranium Miner Oral History and Photography Project, and covers a variety of perspectives, including historical, cultural, and social justice issues.

Burchfield, Larry A. *Radiation Safety: Protection and Management for Homeland Security and Emergency Response.* Hoboken, NJ: Wiley, 2009.

Focuses on radiation detection and response from a homeland security perspective. After an introductory discussion of radiation, nuclear security, and terrorism, the author discusses how ports and other points of entry into the country or into transportation networks can be monitored for the presence of radioactive materials. This is followed by a description of response plans, the role of first responders, treatment of affected persons, and cleanup and decontamination of affected areas.

Committee to Assess Health Risks from Exposure to Low Levels of Ionizing Radiation, National Research Council. *Health Risks from Exposure to Low Levels of Ionizing Radiation: BEIR VII, Phase 2.* Washington, DC: National Academies Press, 2006.

Contains the latest estimates of the risk of cancer and other health effects from exposure to low-level ionizing radiation. Suggests that more accurate figures for cancer and leukemia continue to support the theory that there is no safe lower limit (threshold) for radiation exposure.

Fradkin, Philip L. *Fallout: An American Nuclear Tragedy.* 2nd ed. Boulder, CO: Johnson Books, 2004.

Although the radiation exposure in this case was from nuclear weapons tests in Nevada, not nuclear power accidents, this history of government neglect and legal struggles of victims for redress is an important part of the broader historical question of radiation policy.

Harding, Jim. *Canada's Deadly Secret: Saskatchewan Uranium and the Global Nuclear System*. Halifax, NS: Fernwood, 2007.

A researcher and advocate looks at the lasting effects of uranium mining on people and the environment in one of the world's largest uranium producing regions. Health consequences such as higher cancer risk as well as political issues (including the sovereignty of aboriginal or First Nations peoples) are addressed.

Henriksen, Thormod, and H. David Maillie. *Radiation & Health*. New York: Taylor & Francis, 2003.

A well-organized and clear textbook-style presentation of the characteristics of radiation, how radiation is measured, natural and artificial sources of radiation (including nuclear power plant operation and the fuel cycle), the biological mechanisms behind radiation effects, and radiation in the environment.

Mangano, Joseph J. *Low Level Radiation and Immune System Damage: An Atomic Era Legacy*. Boca Raton, FL: Lewis, 1999.

As a result of numerous aboveground nuclear tests in the 1950s, the human race was subjected to an unplanned "experiment" in exposure to radiation via fallout, plus small amounts from nuclear power plants. This controversial book argues that there is a demonstrable connection between this exposure and the immune system disorders suffered by today's baby boomers.

Mossman, Kenneth L. *Radiation Risks in Perspective*. Boca Raton, FL: CRC / Taylor & Francis, 2007.

Combines an overview of radiation risks with a discussion of peoples' perceptions and attitudes toward risk. Reviews various theoretical approaches and policy responses, suggesting that a distorted perception of risk can lead to unnecessary expenses and unintended consequences.

Ringholz, Raye. *Uranium Frenzy: Saga of the Nuclear West.* Revised and expanded ed. Logan: Utah State University Press, 2002.

With the rush to build a nuclear weapons stockpile and eventually, to fuel the nuclear power industry, the federal government made an all-out effort to procure supplies of uranium. The result was a latter day "gold rush" in Utah and Colorado. This account covers both the investors and fortune seekers who could strike it rich or lose millions of dollars, as well as the uranium miners, many of whom died or got sick as a result of their radiation exposure and who often received little or no compensation.

Schull, William J. *Effects of Atomic Radiation: A Half-Century of Studies from Hiroshima and Nagasaki.* New York: Wiley, 1995.

An analysis of the effects of atomic radiation as assessed over many decades of study of the survivors of Hiroshima and Nagasaki. Includes a detailed chronology of the bombings, the immediate aftermath, and the emergence of longer-term effects.

Stephens, Martha. *The Treatment: The Story of Those Who Died in the Cincinnati Radiation Tests.* Durham, NC: Duke University Press, 2002.

Between 1960 and 1972 the Department of Defense subjected cancer patients to massive radiation doses—sometimes over their whole body. While this was done under the guise of treating their disease, the unknowing patients were actually being used as guinea pigs to determine the effects of radiation on soldiers during a nuclear war. The gripping account includes the struggle to bring these events to public attention and the eventual vindication of surviving family members in court.

Walker, J. Samuel. *Permissible Dose: A History of Radiation Protection in the Twentieth Century.* Berkeley: University of California Press, 2000.

The author, official historian of the Nuclear Regulatory Commission, traces the developing concerns about the effects of radiation and measures to determine what might be an acceptable dose. In addition to radiation from nuclear weapons and power facilities, X-rays and radiation used in medicine have also come under scrutiny. The most detailed focus covers the period from 1950 to 2000. The coverage of regulatory details is probably most useful to specialists.

Yablokov, Alexey V., Vassily B. Nesterenko, and Alexey V. Nesterenko. *Chernobyl: Consequences of the Catastrophe for People and the Environment.* Boston: Blackwell Publishers for the New York Academy of Sciences, 2009.

English translation of an extensive review of the scientific literature that has met with controversy because of its estimate that Chernobyl may have caused up to 1 million deaths. Defenders of the book, however, say that the nuclear establishment and its defenders have greatly minimized death estimates and attributed most of the deaths to causes other than radiation. In turn, many radiation experts have decried the lack of attention to actual radiation doses and reliance on broader ecological and health indicators.

Web Resources

Radiation and Health. New York Department of Health. http://www.health.ny.gov/publications/4402.pdf
Gives basic information on what radiation is, how one is exposed to it, comparison of doses from natural and artificial

sources, and a chart that clarifies the different types of effects and how they relate to doses and thresholds.

Radiation and Your Health. Centers for Disease Control and Prevention.

http://www.cdc.gov/nceh/radiation/
Provides background, including explaining the difference between ionizing radiation (such as from plutonium) and nonionizing radiation, such as from microwaves and cell phones. Includes fact sheets and links such as "What to do during and after a radiation emergency."

Radiation Protection. Environmental Protection Agency.

http://www.epa.gov/radiation/index.html
In addition to providing a variety of background information and resources, the site explains EPA regulations and standards as well as programs to protect water and air, manage waste, and deal with existing contaminated sites.

ToxFAQs for Plutonium. Agency for Toxic Substances and Disease Registry.

http://www.atsdr.cdc.gov/toxfaqs/tf.asp?id=647&tid=119
Summarizes the properties and risks of plutonium and provides answers to frequently asked questions, including how the substance can enter the environment and how it may affect the health of adults and children.

ToxFAQs for Uranium. Agency for Toxic Substances and Disease Registry.

http://www.atsdr.cdc.gov/toxfaqs/tf.asp?id=439&tid=77
Summarizes the properties and risks of uranium and provides answers to frequently asked questions about its health effects. Uranium (unlike plutonium) is a naturally occurring element that is mildly radioactive but can cause health effects if breathed in the form of dust or ingested through drinking water or food. Health effects seem to be primarily chemical rather than radiological. For a

related substance, see the page for Radon at http://www. atsdr.cdc.gov/substances/toxsubstance.asp?toxid=71

Risks, Disasters, and Safety in Nuclear Plants

The works featured here focus on risks and safety issues through the lens of significant nuclear accidents such as Three Mile Island, Chernobyl, and Fukushima. For more general background on radiation risks and effects, see the earlier section "Environmental and Health Effects."

Print Materials

Alexievich, Svetlana, and Keith Gessen. *Voices from Chernobyl.* Normal, IL: Dalkey Archive Press, 2005.

> A collection of accounts by survivors of the 1986 Chernobyl disaster, bringing deep insight into the human dimensions of the tragedy and the actions of the Soviet Union that sometimes made things worse.

Brack, H. G. *Nuclear Information Handbook: A Guide to Accident Terminology and Information Sources.* Liberty, ME: Pennywheel, 2011.

> In response to worldwide concerns brought about by the Fukushima nuclear disaster, this handbook provides both detailed coverage of the effects of this and other major nuclear accidents and a guide to the terminology and units of measurement needed to understand the impact of these events.

Corrice, Leslie. *Fukushima: The First Five Days.* Amazon Digital Services, 2012.

> A good introduction to the design, management, and oversight issues highlighted by the Fukushima nuclear disaster. A bonus is the gripping narrative of plant operators' actual responses to harrowing conditions they faced following the tsunami.

Kingtson, Jeff (ed.). *Natural Disaster and Nuclear Crisis in Japan: Response and Recovery after Japan's 3/11*. New York: Routledge, 2012.

> Contributors explore a broad range of responses involving many aspects of Japanese society, including the media, government agencies, politicians, policy makers, and economists.

Lochbaum, David, Edwin Lyman, Susan G. Stranahan, and the Union of Concerned Scientists. *Fukushima: The Story of a Nuclear Disaster*. New York: New Press, 2014.

> Nuclear experts and journalists combine to offer a vivid, informative, and comprehensive account of the Fukushima events, how and why they happened, official responses, and long term consequences for policy.

Medvedev, Grigori. *The Truth about Chernobyl*. New York: Basic Books, 1991.

> English translation of an account by a nuclear engineer who had worked at Chernobyl. He describes the conditions leading to the accident and the response in the days that followed, including interviews with people who were in fact dying from radiation exposure.

Smith, Gar. *Nuclear Roulette: The Truth about the Most Dangerous Energy Source on Earth*. White River Junction, VT: Chelsea Green, 2012.

> Argues that not only do recent disasters, including Chernobyl and Fukushima, show that nuclear power can have disastrous consequences, but also that inept managers and nuclear apologists in government tends to hide the extent of the risk this technology poses.

Takahashi, Sentaro, ed. *Radiation Monitoring and Dose Estimation of the Fukushima Nuclear Accident*. New York: Springer Verlag, 2014.

Extensive and rather technical survey of the methodology used for estimating the radiation effects associated with the Fukushima meltdowns and subsequent events. Includes useful discussion of how radiation is dispersed into and taken up within the environment.

Web Resources

Adams, Rod. Atomic Insights.

http://atomicinsights.com
A blog that is unabashedly pro-nuclear. Offers news and takes on antinuclear activists on safety issues.

Chernobyl's Legacy: Health, Environmental and Socio-Economic Impacts and Recommendations to the Governments of Belarus, the Russian Federation and Ukraine.

http://www.iaea.org/Publications/Booklets/Chernobyl/
chernobyl.pdf
A publication of the Chernobyl Forum, 2003–2005, with participation by the International Atomic Energy Agency, World Health Organization, and other UN and nongovernmental organizations as well as the affected governments. The report notes that although an increase in thyroid cancer incidence was the only solidly documented health effect of the tragedy, the economic, social, and psychological effects were quite evident and often severe.

Citizens' Nuclear Information Center (Japan).

http://www.cnic.jp/english/
A good source for background information and news about nuclear power in Japan, including fuel cycle and waste issues. The site advocates ending nuclear power and weapons programs, and provides resources and contacts for activists.

"Fukushima and Chernobyl: Myth versus Reality." Argonne National Laboratory, Nuclear Engineering Division.

http://www.ne.anl.gov/jp/fukushima-facts-and-myths.shtml
Mini documentary video assessing and comparing the effects
of Chernobyl and Fukushima. Expert assessments are compared with media exaggerations. Even at Chernobyl, verifi-
able deaths were in the dozens, not tens of thousands—
mainly due to children being fed milk contaminated with
radioactive iodine and developing thyroid cancer. At
Fukushima, no deaths are expected as a result of radiation,
even among workers—but the economic and social disruption is acknowledged to be severe.

Interviews on Situation at Japan's Nuclear Reactors. Argonne
National Laboratory, Nuclear Engineering Division.

http://www.ne.anl.gov/jp/
A collection of links to interviews and multimedia presentations on the impact and significance of the nuclear disaster in Japan. Includes an assessment of the core damage
("worse than feared"), the challenge of cleanup in the area,
and an assessment of radiation exposure.

"Lessons from Fukushima." University of Chicago.

http://nuclear-discussion.uchicago.edu/
Website with presentations from a conference held April 21,
2011, by the University of Chicago Alumni Association,
Argonne National Laboratory, and the Harris Energy
Policy Institute. The panel included Hussein Khalil, director
of the Nuclear Energy Division at the Argonne National
Laboratory and Kennette Benedict, director of the Bulletin
of the Atomic Scientists. In addition to the video of the
event, there are links to related articles and papers.

"Preliminary Dose Estimation from the Nuclear Accident after
the 2011 Great East Japan Earthquake and Tsunami." World
Health Organization, 2012.

http://www.who.int/ionizing_radiation/pub_meet/
fukushima_dose_assessment/en/

In addition to the dose estimates, this report provides maps and illustrations that help visualize the extent and mechanisms by which radiation has spread from the accident. The use and efficacy of protective measures is also assessed.

Special Report: The Fallout from Fukushima. *New Scientist.*
http://www.newscientist.com/special/fukushima-crisis
Links to a variety of articles on the aftermath of the Fukushima disaster and related assessments of the safety of nuclear power (pro and con).

Three Mile Island: The Inside Story. Smithsonian National Museum of American History.
http://americanhistory.si.edu/tmi/index.htm
A companion site for a museum exhibit, this includes a chronological narration and describes how researchers later examined the reactor with cameras and sonar equipment that revealed the true extent of the damage and brought clarity about what had actually happened and how.

Nuclear Waste and the Fuel Cycle

This section brings together materials relating mainly to the life of radioactive materials outside the reactor itself—uranium mining, plutonium processing, and the management of spent fuel and other nuclear waste.

Print Materials

Gephart, Roy E. *Hanford: A Conversation about Nuclear Waste and Cleanup.* Columbus, OH: Battelle, 2003.
The massive Hanford nuclear facility began as the site for production of plutonium from uranium (via nuclear reactors) during the Manhattan Project. Until it was closed in

1990, Hanford produced two-thirds of the plutonium used in the United States. The legacy is many tons of waste stored in tanks and underground facilities. Through this vivid and well-illustrated story of Hanford, the author explains the technical and political issues arising from the need to cope with the environmental problems and risks caused by high-level nuclear waste.

Kazimi, Mujid S., Ernest J. Moniz, and Charles W. Forsberg. *The Future of the Nuclear Fuel Cycle: An Interdisciplinary MIT Study.* Cambridge, MA: MIT Press, 2010.

This report updates two previous MIT studies (2003 and 2009), provides a good introduction to the complexities of the nuclear fuel cycle, and describes challenges that must be met if the United States is to expand nuclear power. These challenges are not only technical, but political and economic—and a postscript following the 2011 Fukushima disaster emphasizes the need to address risks involved with fuel and waste management.

Walker, J. Samuel. *The Road to Yucca Mountain: The Development of Radioactive Waste Policy in the United States.* Berkeley: University of California Press, 2009.

This detailed (and award winning) analysis uses the failure of the decades-long attempt to establish a national nuclear waste repository to look at the way in which scientists, policy makers, legislators, and activists struggled with the waste issue, which remains unresolved today.

Wilson, P. D. (ed.). *The Nuclear Fuel Cycle: From Ore to Waste.* New York: Oxford University Press, 1996.

A comprehensive but not excruciatingly detailed treatment of every step in the life of nuclear fuel: mining and milling, purification, fabrication of fuel assemblies, enrichment, the fission process, reprocessing and fuel recycling, waste management, and environmental effects.

Web Resources

Bunn, Mathew et al. "The Economics of Reprocessing vs. Direct Disposal of Spent Nuclear Fuel." Belfer Center for Science and International Affairs, December 2003.

> http://belfercenter.ksg.harvard.edu/files/repro-report .pdf
> Reprocessing of spent fuel has been advocated as a way to reduce the volume of waste and save costs by producing new fuel. This report, however, uses detailed analysis of capital and operating costs, fuel prices, and other factors to conclude that direct disposal is the better option for both economic and nonproliferation reasons.

Bunn, Matthew et al. "Spent Nuclear Fuel: A Safe, Flexible, and Cost-Effective Near-Term Approach to Spent Fuel Management." Harvard University and University of Tokyo, 2001.

> http://belfercenter.ksg.harvard.edu/files/spentfuel.pdf
> Explores the viability of interim storage of spent fuel. In light of the shutdown of the Yucca Mountain site, this is probably going to be the only viable option for the near to mid term. The authors, representing Harvard University's Project on Managing the Atom and the University of Tokyo's Project on Sociotechnics of Nuclear Energy, explore the technical capabilities, legal and political issues, and public perceptions involved in the siting of waste facilities. The authors advocate a better and more democratic public process and international co-operation in constructing new storage sites.

The Nuclear Fuel Cycle. BBC News.

> http://news.bbc.co.uk/2/shared/spl/hi/sci_nat/05/ nuclear_fuel/html/mining.stm
> A good brief summary of each of the steps of the fuel cycle: mining, conversion, reactor operation, reprocessing, enrichment, and production of nuclear weapons.

Proliferation and Terrorism

An extensive literature exists on the problem of nuclear proliferation, the control of nuclear materials, and efforts to prevent
terrorists from getting their hands on them.

Print Materials

Albright, David. *Peddling Peril: How the Secret Nuclear Trade
Arms America's Enemies*. New York: Free Press, 2010.

> A leading expert on nuclear proliferation recounts his
> investigations and paints a frightening picture of the ease
> with which nuclear materials slip past controls and feed
> black markets. A centerpiece of the trade was Pakistani sci-
> entist and nuclear arms dealer A. Q. Khan, but the author
> says that despite crackdowns in recent years, the trade in
> nuclear technology and materials continues to flourish.

Allison, Graham. *Nuclear Terrorism: the Ultimate Preventable
Catastrophe*. New York: Times Books, 2004.

> The author has impeccable academic credentials (a found-
> ing dean of Harvard's John F. Kennedy School of
> Government) and a distinguished career in government ser-
> vice. This perhaps lends weight to what might be dismissed
> as an overly alarmist depiction of the ease with which terro-
> rists might be able to obtain a "small" nuclear device, smug-
> gle it into the United States, and destroy the heart of a great
> city such as New York or San Francisco. The author does
> suggest that a difficult but practical path to preventing such
> attacks is to enforce a world of "Three No's"—no loose
> nukes, no "nascent nukes," and no new nuclear states.

Babkina, A. M. *Nuclear Proliferation: A Bibliography with
Indexes*. 2nd ed. New York: Nova Science Publishers, 2002.

> No longer current but useful historically for examining
> the state of the debate following the collapse of the
> Soviet Union and early concerns about nuclear terrorism.

Bahgat, Gawdat G. *Proliferation of Nuclear Weapons in the Middle East*. Gainesville: University Press of Florida, 2008.

>A good complement to the many "U.S.-centric" and Cold War-influenced analyses of nuclear proliferation, this treatment gives full consideration to the historical and economic motivations that have impelled countries in the region to seek nuclear capability.

Blackwill, Robert D. (ed.). *Iran: The Nuclear Challenge*. New York: Council on Foreign Relations, 2012.

>A group of experts under the auspices of the Council on Foreign Relations look at a number of possibilities. What if negotiations break down? What might be the result of a U.S. attack on Iran? Could the world live with a nuclear-armed Iran?

Brown, Michael E. et al. *Going Nuclear: Nuclear Proliferation and International Security in the 21st Century*. Cambridge, MA: MIT Press, 2010.

>A collection of essays exploring the causes and potential consequences of nuclear proliferation. The authors examine issues such as the motives of nations in seeking nuclear capability, case studies (including nations that decided to give up the bomb), and the challenge of nuclear terrorism.

Campbell, Kurt M., Robert J. Einhorn, and Mitchell Reiss (eds.). *The Nuclear Tipping Point: Why States Reconsider Their Nuclear Choices*. Washington, DC: Brookings Institution Press, 2004.

>A wide-ranging collection of essays that examine the history of nuclear proliferation and attempts to restrain it. There are case studies of individual countries such as Egypt, Syria, Saudi Arabia, Japan, South Korea, and Taiwan that evaluate their nuclear decisions and possible future moves. The discussion is influenced by the concept of a "tipping point"—factors that may lead a number of nations to move decisively toward acquiring nuclear capability.

Caravelli, Jack. *Beyond Sand and Oil: The Nuclear Middle East.* Santa Barbara, CA: Praeger, 2011.

A detailed overview of the nuclear activities of countries in the Middle East and surrounding areas—Iran, Iraq, Israel, Libya, Pakistan, and India. There is a particular focus on Iran's nuclear program and reactions to it by Israel and the United States, as well as potential instability and Islamic insurgency in nuclear-armed Pakistan.

Caravelli, Jack. *Nuclear Insecurity: Understanding the Threat from Rogue Nations and Terrorists.* Santa Barbara, CA: Praeger, 2007.

Having personally played an important role in securing nuclear materials in the late 1990s, the author offers a sobering assessment of the flaws in the effort to safely contain several hundred tons of highly enriched nuclear material in Russia and elsewhere.

Corera, Gordon. *Shopping for Bombs: Nuclear Proliferation, Global Insecurity, and the Rise and Fall of the A. Q. Khan Network.* New York: Oxford University Press, 2006.

A. Q. Khan, a top Pakistani nuclear scientist, played a key role in developing Pakistan's uranium enrichment program and thus the country's nuclear arsenal. However, at the same time Khan was developing contacts that brought technical assistance to China's nuclear program in the 1980s and allegedly to North Korea, Libya, and Iran in the 1990s. The author describes how Khan developed his network and efforts by the International Atomic Energy Agency along with U.S. and other Western government agencies to dismantle it.

Doyle, James E. et al. *Nuclear Safeguards, Security, and Nonproliferation: Achieving Security with Technology and Policy.* Boston: Butterworth-Heinemann, 2008

A collection of essays and case studies involving many countries, exploring just about every aspect of securing and monitoring nuclear materials around the world.

Ferguson, Charles D., and William C. Parker. *The Four Faces of Nuclear Terrorism*. New York: Routledge, 2005.

> Discusses step by step just what is involved in nuclear terrorism. Who are the terrorists, and what are their capabilities? How might terrorists be able to seize a ready-made nuclear weapon or fabricate their own? What are the vulnerabilities of nuclear power plants with their stockpiles of nuclear fuel? There is also a discussion of dirty bombs and other "radiological devices" as well as "a plan for urgent action against nuclear terrorism."

Frantz, Douglas, and Catherine Collins. *The Man from Pakistan: The True Story of the World's Most Dangerous Nuclear Smuggler*. New York: Twelve, 2008.

> An interesting account of master nuclear proliferator A.Q. Khan and the intricate Pakistani and international political and diplomatic web in which he worked. The authors, both journalists who base their account on numerous interviews and government reports, point out a number of ways in which Western governments and antiproliferation agencies could have detected or blocked Kahn's activities.

Jenkins, Brian Michael. *Will Terrorists Go Nuclear?* Amherst, NY: Prometheus, 2008.

> A leading terrorism expert provides an assessment of the actual capabilities and motives of groups such as Al Qaeda, as well as the extent of a black market in nuclear materials and consideration of the technical challenges faced by would-be bomb makers. The author's ultimate concern, however, is what he sees to be the fear and exaggeration of the threat and the danger such unreasonable beliefs pose to domestic civil liberties and international relations.

Joyner, Daniel H. *Interpreting the Nuclear Non-Proliferation Treaty*. New York: Oxford University Press, 2013.

The author presents a detailed analysis and argument that the nonproliferation aspects of the treaty have been emphasized at the expense of what had been the important intention to foster the development of peaceful nuclear power. The result has been an unsound legal framework.

Khan, Saira. *Iran and Nuclear Weapons: Protracted Conflict and Proliferation*. New York: Routledge, 2009.

Uses Iran as a case study of proliferation in the context of protracted conflict and the very different perceptions and motivations of the parties involved. In particular, the asymmetry in relation to a global nuclear power (the United States) is suggested to be a driving force toward proliferation.

Maddock, Shane J. *Nuclear Apartheid: The Quest for American Atomic Supremacy from World War II to the Present*. Chapel Hill: University of North Carolina Press, 2010.

A provocative argument that an ideology that only certain nations (particularly the United States) could be trusted with nuclear weapons led to the division of the world into nuclear haves and have-nots—and led to a great desire on the part of China, India, and later other nations to join the "nuclear club."

Mozley, Robert F. *The Politics and Technology of Nuclear Proliferation*. Seattle: University of Washington Press, 2000.

Although some details are outdated, this book is still valuable for the clear and detailed explanation of how nuclear materials are processed and the knowledge involved in constructing nuclear weapons, as well as approaches to preventing proliferation.

Mueller, John E. *Atomic Obsession: Nuclear Alarmism from Hiroshima to Al-Qaeda*. New York: Oxford University Press, 2010.

This is a useful counterpoint to the many books and articles that assert that nuclear terrorism is a clear and present danger that requires urgent action. Using a broader context of nuclear fears versus reality, the author suggests that the effects of nuclear weapons on history have been overstated, proliferation has been slow because the perceived value of nuclear weapons has been low, and the capabilities of would-be nuclear terrorists as well as their desire to acquire such weapons have also been overstated.

Reed, Thomas C., and Danny B. Stillman. *The Nuclear Express: A Political History of the Bomb and Its Proliferation.* Minneapolis: Zenith, 2009.

One of the more accessible and lively accounts of how nuclear weapons spread from being a monopoly of the United States and then just a handful of countries . . . to a proliferation that today seems to be out of control.

Sagan, Scott D., and Kenneth N. Waltz. *The Spread of Nuclear Weapons: An Enduring Debate.* 3rd ed. New York: W. W. Norton, 2013.

The latest revision of an overview on nuclear proliferation that has been widely used in international relations courses. The new edition includes expanded and updated discussion of North Korea, Iran, and Iraq.

Spellman, Frank R., and Melissa L. Stoudt. *Nuclear Infrastructure Protection and Homeland Security.* Lanham, MD: Government Institutes, 2011.

Although primarily intended for managers and other employees at nuclear installations, this guide provides a useful and rather detailed overview of the types of threats that must be countered with regard to nuclear power facilities and the transport, handling, and storage of radioactive materials and waste.

Weiner, Sharon K. *Our Own Worst Enemy? Institutional Interests and the Proliferation of Nuclear Weapons Expertise.* Cambridge, MA: MIT Press, 2011.

> The author, a professor of international studies, argues that despite a lot of goodwill on both sides, the attempt to limit the proliferation of nuclear knowledge following the collapse of the Soviet Union did not succeed. The various U.S. agencies involved come in for particular criticism.

Wirtz, James J., and Peter R. Lavoy (eds.). *Over the Horizon Proliferation Threats.* Stanford, CA: Stanford University Press, 2012.

> Most discussion of nuclear proliferation "hot spots" has focused on countries such as Iran, Iraq, Pakistan, and North Korea. The contributors to this collection explore the current situation and proliferation risks involving Japan, Taiwan, Saudi Arabia, South Africa, Burma, and countries in Southeast Asia. Other contributors look at instructive cases such as South America (Argentina, Brazil, and Venezuela) that turned away from the nuclear path, as well as Ukraine, which "inherited" a nuclear arsenal from the Soviet Union.

Web Resources

Alger, Justin. "A Guide to Global Nuclear Governance: Safety, Security and Nonproliferation." Center for International Governance Innovation, September 2008.

> http://www.cigionline.org/sites/default/files/A%20Guide%20to%20Nuclear%20Power.pdf
> Summarizes the treaties, agreements, arrangements, and initiatives involving nuclear technologies and materials.

Bronner, Michael. "100 Grams (and Counting . . .): Notes from the Nuclear Underworld." Belfer Center for Science and International Affairs, Harvard Kennedy School, June 2008.

http://belfercenter.ksg.harvard.edu/files/100-Grams-Final
-Color.pdf
Most discussion of nuclear proliferation seems to be
abstract or hypothetical, but here a journalist takes readers
into the circumstances of the actual smuggling of highly
enriched uranium (HEU) in the republic of Georgia. In
2006 the smugglers were discovered and the uranium
seized, but both government authorities and the smugglers
themselves seemed not to be very competent. The author
emphasizes the importance of vigilance and effective monitoring, as in dealing with other types of smuggling.

Global Governance Monitor: Nuclear Proliferation. Council
on Foreign Relations.
http://www.cfr.org/global-governance/global-governance
-monitor/p18985?breadcrumb=%2Fthinktank%2Fiigg%
2Fpublications#/Nonproliferation/
Provides a video overview, timeline, brief on current
issues, interactive maps, and a resource guide.

Nuclear Proliferation. *Economist*.
http://www.economist.com/topics/nuclear-proliferation
A topical portal linking to articles on nuclear proliferation,
weapons programs, and related issues around the world.

New Designs and Technologies

A number of new reactor designs have been in development.
This section deals with works that describe this new technology
and its advantages, as well as those that take a more critical
view. Some works on fusion energy research are also included.

Print Materials

Cochran, Thomas et al. *Fast Breeder Reactor Programs: History and
Status*. Princeton, NJ: International Panel on Fissile Materials, 2010.

Fast breeder reactors offer the capability of producing new fuel by converting "fertile" isotopes to fissionable plutonium. The authors explore the history of this technology in the United States, Russia, India, Japan, and Britain and the reasons it so far has not become economically or politically viable, including technical challenges, environmental risks, and proliferation issues.

Singer, Neal. *Wonders of Nuclear Fusion: Creating an Ultimate Energy Source.* Albuquerque: University of New Mexico Press, 2011.

This overview of nuclear fusion and account of the development of experimental fusion devices is written for young readers but is equally useful for older students and adults.

Till, Charles E., and Yoon Il Chang. *Plentiful Energy: The Story of the Integral Fast Reactor: The Complex History of a Simple Reactor Technology, with Emphasis on its Scientific Bases for Non-Specialists.* CreateSpace (Amazon), 2011.

Two leading developers of the integral fast reactor at the Argonne National Laboratory in the 1980s and early 1990s argue that it revolutionizes the nuclear fuel cycle by safely recycling and generating new fuel, largely eliminating the problem of spent fuel disposal. There is also insight into the management of and challenges faced by such complex research programs.

Web Resources

Advanced Nuclear Power Reactors. World Nuclear Association. http://www.world-nuclear.org/info/Nuclear-Fuel-Cycle/ Power-Reactors/Advanced-Nuclear-Power-Reactors/ Summarizes new nuclear reactor designs being developed in several countries, and discusses the characteristics and advantages of these systems.

Adee, Sally, and Erico Guizzo. "Nuclear Reactor Renaissance." *IEEE Spectrum*, August 2010.

> http://spectrum.ieee.org/energy/nuclear/nuclear-reactor
> -renaissance/0
> Gives an overview of the types of next-generation reactors (advanced light water reactors, small modular reactors, and Generation IV reactors) and surveys the following designs: Westinghouse AP-1000, EPR (Evolutionary Power Reactor), NucScale, Hyperion Power Module, Toshiba 4S, Next Generation Nuclear Plant, and TerraPower TP-1.

Waldrop, M. Mitchell. "Nuclear Energy: Radical Reactors." *Nature,* December 5, 2012.

> http://www.nature.com/news/nuclear-energy-radical-reactors
> -1.11957
> Suggests that the molten salt reactor and other new technologies may have substantial advantages over the existing (and now almost venerable) light water reactors using conventional fuel rods. A historical overview explains why the existing technology was chosen and why once promising options such as fuel processing were not used (due to proliferation concerns). The current U.S. Nuclear Power 2010 program of the U.S. DOE is reviewed, along with current designs in development and potential investors.

Climate Change and the Future of Nuclear Power

The debate over whether to expand or phase out nuclear power is now focused on whether this technology can provide the means needed to move away from fossil fuels and their carbon dioxide emissions.

Print Materials

Brook, Barry W., and Ian Lowe. *Why vs. Why: Nuclear Power.* Neutral Bay, Australia: Pantera, 2010.

In this succinct but valuable book, two leading environmental and climate scientists take opposing positions on whether nuclear power should be expanded to cope with the ongoing climate crisis. Brook argues that the safety and security of nuclear power can be strengthened and the industry can lead a new "clean energy revolution." Lowe counters that the problems with nuclear power cannot be solved easily, and that at any rate nuclear is too expensive and slow an option to be useful in dealing with climate change.

Cadenas, Juan José Gomez. *The Nuclear Environmentalist: Is There a Green Road to Nuclear Energy?* New York: Springer, 2012.

Attempts to dispel misconceptions about radioactivity and nuclear reactors and argues that nuclear power has a place in a rational discussion of the various alternatives to fossil fuels.

Caldicott, Helen. *Nuclear Power Is Not the Answer*. New York: New Press, 2007.

The renowned (and controversial) physician-activist takes on the new arguments in favor of a "nuclear renaissance." She argues the technology is still too risky and expensive, no solution to the waste problem is in sight, and safer energy alternatives are readily available.

Cochran, Thomas B. et al. *Position Paper: Commercial Nuclear Power*. New York: National Resources Defense Council, 2005.

The NRDC addresses claims that nuclear power should be used to combat climate change. Perhaps not surprisingly, the report concludes that economic liabilities, environmental impact, accident risks, and proliferation and security issues combine to make nuclear power a poor choice for the indefinite future.

Cravens, Gwyneth, and Richard Rhodes. *Power to Save the World: The Truth about Nuclear Energy*. New York: Vintage, 2008.

A science reporter delves into nuclear power and finds skepticism yielding to a conviction that the energy source can be used safely and responsibility, with less serious environmental effects than the continued use of fossil fuels for the majority of power generation.

Domenici, Pete V. *A Brighter Tomorrow: Fulfilling the Promise of Nuclear Energy.* Lanham, MD: Rowman & Littlefield, 2007.

The senior senator from New Mexico and an acknowledged congressional energy expert, Domenici delivers a strong, wide-ranging case for expanding nuclear power in the early twenty-first century. He proposes an interim solution to break the impasse over waste storage, arguing the future technology will bring better long-term solutions.

Elliott, David (ed.). *Nuclear or Not? Does Nuclear Power Have a Place in a Sustainable Energy Future?* New York: Palgrave Macmillan, 2007.

An overview and collection of views pro and con concerning the role that should be played by nuclear energy in determining the future energy mix. Topics covered include economics of nuclear power, nuclear versus sustainable energy, radiation and waste problems, risks, and insecurity. A British perspective that may broaden a discussion that has been dominated largely by American voices.

Hore-Lacy, Ian. *Nuclear Energy in the 21st Century: World Nuclear University Primer.* 3rd revised ed. London: World Nuclear University Press, 2012.

Provides a good factual background and discussion of nuclear power and related issues, and answers to specific concerns. Generally argues in favor of the further development and expansion of nuclear power.

Makhijani, Arjun. *Carbon-Free and Nuclear Free: A Roadmap for U.S. Energy Policy.* Muskegon, MI: RDR Books and Takoma Park, MD: IEER Press, 2007.

> An expert on energy economics and conservation offers the results of a systematic study leading to a phased plan for moving to a carbon-free energy grid without assuming the risks of proliferation and other problems associated with nuclear power.

Palley, Reese. *The Answer: Why Only Inherently Safe, Mini Nuclear Power Plants Can Save Our World.* New York: Quantuck Lane, 2011.

> Critiques alternatives to fossil fuels, suggesting that they often actually have a comparable carbon impact. Argues that self-contained, inherently safe nuclear reactors can provide a versatile local source of clean power for homes and industry.

Smith, Brice. *Insurmountable Risks: The Dangers of Using Nuclear Power to Combat Global Climate Change.* Muskegon, MI: RDR Books, 2006.

> A head-on challenge to the arguments for nuclear power as a choice to replace fossil fuels and their carbon burden. The author argues that the cost and risks of nuclear power are unacceptable and that there are alternative technologies that can adequately reduce carbon emissions.

Tucker, William. *Terrestrial Energy: How Nuclear Power Will Lead the Green Revolution and End America's Energy Odyssey.* Washington, DC: Bartleby, 2008

> A veteran journalist investigates the history and development of all major forms of energy. He then turns to the story of nuclear energy and argues that the nuclear industry has been learning from its mistakes and now seeks new

opportunities in the twenty-first century. It is suggested that while various forms of alternative energy can make important contributions, only nuclear can provide sufficient carbon-free energy to power the widespread use of electric vehicles.

Web Resources

CASEnergy Coalition. "Clean and Safe Energy."
    http://casenergy.org/
    Pro-nuclear site belonging to a coalition of about 900 organizations involved with nuclear power. Includes news, basic resources, and a blog.

Just the Facts: A Look at the Five Fatal Flaws of Nuclear Power. Public Citizen.
    http://www.citizen.org/cmep/article_redirect.cfm?ID=13447
    A collection of fact sheets on the cost, security, safety, waste, and proliferation involved with nuclear power, plus a summary. The overall view is that in all of these areas nuclear power fails to be a safe or viable option for meeting future energy needs.

"Letter to President Obama." Nuclear Information and Resource Service.
    www.nirs.org/nukerelapse/obama/finalcesletter.pdf
    A letter of January 24, 2011, urging the president to turn the nation away from nuclear and fossil energy sources and embrace energy efficiency and renewable energy. Signatory groups include Friends of the Earth, Greenpeace USA, Public Citizen, and Beyond Nuclear.

Nuclear Energy and Climate Change. World Nuclear Association.
    http://www.world-nuclear.org/Features/Climate-Change/Climate-Change-and-Nuclear-Energy/

Links to background information, arguments, reports, and news supporting the expansion of nuclear power as a way to combat climate change.

"Nuclear Energy Is Dirty Energy (and does not fit into a "clean energy standard")." Nuclear Information and Resource Service.
http://www.nirs.org/factsheets/
nuclearenergyisdirtyenergy.pdf
A briefing paper that argues that the invisible pollution of radioactivity is at least as harmful as CO_2 or visible pollutants. Describes pollution and environmental effects of the nuclear fuel cycle along with the risks of nuclear accidents, and urges that nuclear power not be included in any clean energy standard.

Schneider, Mycle, Anthony Froggatt, and Steve Thomas. *Nuclear Power in a Post-Fukushima World: The World Nuclear Industry Status Report, 2010–2011; 25 Years after the Chernobyl Accident.* Washington, DC: Worldwatch Institute, 2011.
http://www.worldwatch.org/system/files/pdf/WorldNuclear
IndustryStatusReport2011_%20FINAL.pdf
Includes a summary of the status of nuclear power plants around the world (operating, planned, or being constructed) and a country-by-country report on the status of nuclear programs. Data and trends on the economics of nuclear power are used to support an overall conclusion that expanding nuclear power is not a good option even in the face of climate change.

Following is a chronology of significant events in the development of nuclear power, including technical advances, accidents, and legislation. The chronology runs from the late nineteenth century to the present, with an emphasis on developments in the 1970s and later. Researchers seeking further historical information should consult some of the historical or archival websites described in Chapter 6.

1895 Wilhelm Roentgen, a German physicist, discovers a new form of short-wavelength radiation that can penetrate objects. He calls them X-rays and receives the Nobel Prize in 1901 for his discovery. Besides being very useful to doctors, X-rays will provide a tool that physicists can use to explore the structure of atoms.

1896 French physicist Antoine Henry Becquerel tries to discover whether uranium produces X-rays when exposed to sunlight. After several foggy days he develops the film anyway and is surprised to find that the uranium produces radiation by itself without any need of sunlight. This discovery of naturally

---

A consortium of nuclear power companies wanted to store 40,000 tons of high level nuclear waste at the Skull Valley Goshute Reservation in Nevada. Native Americans were joined by many activists in opposing the waste plan, which was canceled in 2013. (AP Photo/Douglas C. Pizac)

radioactive materials is the first hint of the power locked inside the atom.

1897   British physicist J. J. Thompson discovers the electron, one of three basic particles that make up the atom.

1898   Pierre and Marie Sklodowska Curie, an intrepid husband-and-wife science team, sift through uranium ore, treat it chemically, and discover two radioactive elements, polonium and radium. Radium, a powerful radiation source, is soon used for both medical treatments and research into atomic physics.

1900   French physicist Paul Villard discovers gamma rays, an energetic, deeply penetrating form of radiation.

1905   Albert Einstein introduces his theory of relativity. One consequence of Einstein's work is the equivalence of matter and energy, shown in the famous equation $E = MC^2$. This equation shows why the breakdown of small amounts of matter in the atom can lead to a large release of energy.

1908   New Zealand–born British physicist Ernest Rutherford and his assistant Frederick Soddy receive the Nobel Prize for their explanation of alpha and beta radiation and their demonstration that radioactive emission can change one kind of atom into another (transmutation).

1914   H. G. Wells's novel *The World Set Free* foresees a future where atomic energy is used for both peaceful power generation and devastating weapons.

1932   James Chadwick, a former student of Rutherford's, discovers the neutron, an uncharged particle in the nucleus of the atom that plays a key role in nuclear fission. He receives the Nobel Prize in 1935.

1934 Italian physicist Enrico Fermi bombards uranium atoms with neutrons. Some of the atoms absorb the neutrons and become neptunium, the first artificial element.

1938 German chemists Otto Hahn and Fritz Strassmann bombard uranium with neutrons and are surprised to find they have produced barium and krypton, two much lighter elements. Austrian physicist Lise Meitner and her nephew Otto Frisch discuss these results and conclude that the uranium atom had been actually split in two by the neutron. They apply the biological term "fission" to this process.

1939 Enrico Fermi carries the news of uranium fission to Washington, DC, and predicts the possibility of a chain reaction. Albert Einstein, responding to concern within the scientific community, writes a letter to President Franklin D. Roosevelt, telling him of the possibility of building a weapon of unprecedented power—an atomic bomb.

1940 Researchers at the Radiation Laboratory at the University of California, Berkeley, create plutonium, an artificial element that can be used as a nuclear fuel or to make nuclear weapons.

1941 President Roosevelt gives approval for a secret scientific team headed by Enrico Fermi, J. Robert Oppenheimer, and Ernest O. Lawrence to begin construction of an atomic bomb. The massive project is given the code name Manhattan District Project.

1942 Researchers build the world's first nuclear reactor under the football stands at the University of Chicago.

1943 The Los Alamos Scientific Laboratory is built in New Mexico, and scientists and technicians begin the design and assembly of the first atomic bomb. The government also acquires land in the Colorado Plateau and begins to contract out for uranium mining operations.

1945   The atomic bomb is successfully tested at Alamogordo, New Mexico. The United States drops a uranium-fueled atomic bomb on Hiroshima, Japan, with devastating effects. A second atomic bomb, fueled by plutonium, is dropped on the city of Nagasaki. Japan surrenders unconditionally, ending World War II.

1946   President Harry S. Truman signs the Atomic Energy Act of 1946, establishing the Atomic Energy Commission (AEC). One of the responsibilities of this body is to promote the peaceful use of atomic energy, such as the design and construction of nuclear power plants. The Joint Committee on Atomic Energy (JCAE) is also established.

1947   The Atomic Energy Commission establishes the Reactor Safeguards Committee. The AEC also establishes the Industrial Advisory Group under Chairman James W. Parker to investigate peaceful uses of atomic energy.

1951   The Experimental Breeder Reactor 1 in Arco, Idaho, generates the first electricity from nuclear energy. A breeder reactor, the device also produces its own fuel.

1952   The National Research Experimental Reactor at Chalk River, Canada, gets out of control, leading to a partial core meltdown that is contained.

1953   President Dwight D. Eisenhower delivers what becomes known as his "Atoms for Peace" speech before delegates of the United Nations in New York. He calls for an international agreement to share in the development of nuclear technology.

1954 *Nautilus,* the first nuclear-powered submarine, is launched. Its light water reactor, built under the leadership of Admiral Hyman G. Rickover, demonstrates the feasibility of nuclear power and will become the basic design used by most nuclear power plants.

1954 The Atomic Energy Act of 1954 is passed, revising the 1946 act. The new legislation allows for private ownership of nuclear power facilities and encourages the participation of private industry in the general development and application of nuclear energy. The JCAE approves a five-year program for reactor development, and construction of the first commercial nuclear power plant is begun at Shippingport, near Pittsburgh, Pennsylvania.

1955 The Atomic Energy Commission begins the Cooperative Power Demonstration Program, which offers technical assistance and a reduction of some regulatory fees to any utility that desires to build a nuclear power plant.

1955 The Experimental Breeder Reactor begins operation in Idaho, demonstrating the ability to create not only nuclear power, but also its own fuel.

1955 The first UN International Conference on the Peaceful Uses of Atomic Energy is held at Geneva, Switzerland.

1956 Development of power reactors is well underway in Great Britain, France, and the Soviet Union.

1957 The Brookhaven Report, "Theoretical Possibilities and Consequences of Major Accidents in Large Nuclear Plants," is released. It seeks to reassure the public and industry officials that most anxiety about nuclear power is unfounded. At the same time, however, a secret report by the same organization estimates that a nuclear meltdown could cause $7 billion in damage, kill several thousand people, and contaminate 50,000 square miles of land.

1957   President Eisenhower signs the Price-Anderson Act, which encourages utilities to invest in nuclear power plants. The law limits the financial liability of plant owners and contractors in the event of a major accident.

1957   The International Atomic Energy Agency (IAEA) is established to promote and monitor the peaceful uses of nuclear energy throughout the world.

1957   The first U.S. commercial nuclear power station, at Shippingport, Pennsylvania, begins operation. This demonstration 60-megawatt plant uses a pressurized water reactor and is soon supplying electric power to the surrounding community.

1958   The European Nuclear Energy Agency (later the Nuclear Energy Agency) is founded.

1960   The first boiling water reactor, manufactured by General Electric, goes into service with the opening of Commonwealth Edison's Dresden Plant 1.

1961   A test reactor in Idaho Falls, Idaho, goes out of control, resulting in the death of three maintenance workers.

1964   President Lyndon B. Johnson signs the Private Ownership of Special Nuclear Materials Act, which allows the power industry to own fuel for power reactors (rather than leasing it from the government). This allows a more normal market to develop.

1965   Expressing growing confidence in the nuclear market, the big reactor manufacturers General Electric, Babcock and Wilcox, Westinghouse, and Combustion Engineering offer their products at a fixed cost (plus inflation allowance). Between 1965 and 1967, utilities order 50 nuclear plants with a total of 40,000 megawatts of generating capacity.

1966   The Fermi fast breeder reactor outside of Detroit, Michigan, suffers a partial core meltdown that is concealed from the public. The safety of the reactor had been challenged by the AFL-CIO and other critics, but the U.S. Supreme Court had overturned a U.S. Court of Appeals decision that had halted construction of the reactor in 1956.

1969   India's first nuclear power plant, the Tarapur atomic power station, goes online. It was built by General Electric.

1969   The AEC establishes the Atomic Safety and Licensing Appeal Board, whose three members are responsible for technical review of the safety systems and procedures required for the granting of a construction permit for a nuclear facility.

1973   The Organization of Petroleum Exporting Countries (OPEC) begins an oil embargo that leads to an "energy crisis" in the United States. Nuclear advocates quickly point to the expanded development of nuclear power as a way to lessen dependence on foreign energy supplies.

1974   The Energy Reorganization Act is signed by President Gerald Ford. The act replaces the Atomic Energy Commission with two new agencies, the Energy Research and Development Administration (ERDA) and the Nuclear Regulatory Commission (NRC).

1974   Daniel Ford and Henry Kendall, representing the Union of Concerned Scientists and the Friends of the Earth, publish a report that concludes that emergency core cooling systems in nuclear reactors may not work.

1975   The NRC orders the shutdown of 23 nuclear reactors because of cracking in the coolant pipes.

1975   A fire started by a candle being used to find leaks damages the core at the Brown's Ferry nuclear plant in Decatur, Alabama.

1976   Nuclear engineers Gregory C. Minor, Richard B. Hubbard, and Dale G. Bridenbaugh resign their positions and "blow the whistle" to journalists on what they consider to be lax safety procedures at plants operated by General Electric's nuclear safety division.

1976   The Friends of the Earth and other organizations issue a report that views nuclear power not from the standpoint of safety, but from a concern that the centralized political power of a nuclear infrastructure is antithetical to democratic values.

1977   President Jimmy Carter announces that the United States will postpone indefinitely its plans to reprocess spent nuclear fuel. He also proposes to terminate the Clinch River breeder reactor in Virginia.

1977   President Carter signs the Energy Reorganization Act, which creates the Department of Energy and combines the Energy Research and Development Administration and the Federal Energy Administration.

1977   The Carolina Environmental Study Group, a group of citizen activists, enlists the aid of the Public Citizen Litigation Group to challenge the Price-Anderson Act as being unconstitutional. In June, a federal district court agrees, saying that the law violates the rights of victims of nuclear accidents by denying them the just compensation required under the Fifth Amendment. (See *Carolina Environmental Study Group vs. Atomic Energy Commission*, 431 F. Supp. 230, W.D.N.C. 1977.)

1978   The U.S. Supreme Court overturns the federal district court decision of the previous year, preserving the nuclear industry's limited liability.

1979   A stuck valve and confusion on the part of operators lead to an accident at the Three Mile Island nuclear power plant near Harrisburg, Pennsylvania. It is the most serious accident to a commercial nuclear plant in U.S. history. Although no one is directly injured by the accident, radioactive material is released into the atmosphere.

1982   The Nuclear Waste Policy Act (NWPA) begins a process of environmental assessment that is supposed to result in the choice of a site for a permanent, national, high-level waste storage facility. The process will encounter numerous setbacks.

1983   The Senate refuses further funding of the Clinch River breeder reactor, effectively closing down the project.

1986   The world's worst nuclear power plant accident takes place at the Soviet nuclear facility at Chernobyl in the Ukraine. An explosion destroys the plant, and 35 plant workers are killed. A large amount of radiation is released into the environment, about 1 million times what had been released in the Three Mile Island accident. There are more than 200 cases of serious radiation sickness, and exposure is likely to cause dozens of cases of leukemia and other cancers in the future.

1986   The Nuclear Regulatory Commission reports that 430 nuclear plant shutdowns due to emergency have taken place.

1986   The NRC cites 492 nuclear safety violations at U.S. power reactors.

1988 The DOE awards the Bechtel Group, Inc., a large engineering and mining firm, a $1 billion contract to develop a system to transport and store radioactive nuclear wastes.

1990 An almost-completed nuclear power station in Midland, Michigan, is converted to burning natural gas after its owners are convinced that widespread political opposition and economic factors make nuclear operation too expensive.

1991 Japan begins construction of an advanced boiling water reactor (ABWR). A consortium of General Electric, Hitachi, and Toshiba successfully markets the new design in Asia.

1991 The International Commission on Radiation Protection accepts the theory that any dose of radiation, however small, is harmful.

1994 Princeton University's donut-shaped Tokamak fusion reactor produces 10.7 megawatts of power, setting a new record but not reaching commercial viability.

1994 At the end of the year, 424 nuclear reactors are operating worldwide. The fastest growth is in Asia, where 14 reactors are under construction. Japan has 48 nuclear plants that produce 30 percent of its electrical energy.

1996 By the end of this year, six nuclear power plants close in the United States because they are no longer economical to operate. Meanwhile, Canada closes 21 nuclear plants for safety reasons.

1996 The Tokyo Electric Power Company (TEPCO) begins operating the first advanced boiling water reactor (ABWR). By the next year TEPCO is operating the world's largest reactor complex.

1997   The Earth Summit at Kyoto, Japan, includes a proposal for nations to reduce their greenhouse gas emissions 7 percent below 1990 levels by 2012. Nuclear power advocates suggest that expanding use of nonburning nuclear power is a key to reducing the need for "dirty" fossil fuels.

1998   The Nuclear Regulatory Commission certifies several advanced nuclear power plant designs, allowing builders of new plants to obtain building and operating licenses in a single step before beginning construction.

1998   The Union of Concerned Scientists reports that their 15-month study of 10 nuclear plants reveals careless inspections, numerous worker errors, and faulty procedures.

1998   Workers at a Japanese uranium processing plant accidentally set off a runaway nuclear reaction. Two of the three workers will die from the effects of massive radiation exposure.

1998   A U.S. federal appeals court reinstates 2,000 lawsuits claiming health damages from the Three Mile Island accident. The judge rules that the lower court made an error in dismissing the suits on the basis of a sample of 10 cases.

2000   Steam generation tubes rupture at the Indian Point 2 Nuclear Reactor in Buchanan, New York. Although only a small amount of radioactive steam is released into the atmosphere, critics such as Ralph Nader's Public Citizen group warn that the rupture of similar tubes in aging nuclear power plants could cause disastrous core meltdowns.

2000   A 13-year study by the University of Pittsburgh finds no increased incidence of cancer in people living near the Three Mile Island nuclear plant.

2000   German chancellor Gerhard Schroeder and the nuclear industry agree to phase out the country's nuclear power plants over a 20-year period.

2001   The September 11 attacks heighten concerns about terrorist attacks on nuclear facilities and possible terrorist use of nuclear weapons or materials in future attacks.

2002   The discovery of a badly corroded reactor vessel head at the Davis-Besse nuclear plant raise questions about safety inspections and leads to NRC sanctions and criminal indictments.

2003   Alleged Iraqi progress toward developing nuclear weapons is one of the reasons given for the United States invading Iraq and overthrowing Saddam Hussein.

2004   Concern about nuclear proliferation grows when Abdul Khan of Pakistan admits to selling nuclear technology to Iran, Libya, and North Korea.

2005   Lithuania, a country heavily dependent on nuclear power, decides to shut down half of its nuclear generating capacity, which was provided by reactors using the same design as those at Chernobyl. Meanwhile, Finland approves construction of a large nuclear plant.

2006   North Korea tests its first nuclear device. Further tests of nuclear devices and longer-range missiles continue despite vigorous opposition from the United States and other countries.

2007   The Nuclear Regulatory Commission receives its first application to build a new nuclear power plant since 1979. Meanwhile, the Browns Ferry Unit 1 plant, which had been shut down in 1985, is brought back online.

2010   The Obama administration quietly abandons the decades-long attempt to establish Yucca Mountain as the nation's main nuclear waste depository.

2010   In connection with a Nuclear Security Summit, President Obama announces a goal of securing all the world's vulnerable nuclear materials within four years.

2011   In March a major earthquake and the resulting tsunami flood and disable several reactors in Fukushima, Japan, leading to a widespread release of radioactivity and a widening evacuation. In response, Germany says it will phase out its nuclear industry, and most Japanese plants would remain shut down for months or possibly permanently.

2011   In July the Blue Ribbon Commission on America's Nuclear Future proposes establishing centralized interim facilities to store high-level nuclear waste until a permanent facility can be established.

2012   All commercial reactors in Japan remain shut down.

2012   In March, The White House announces that $450 million will be made available to support efforts to design and certify small modular nuclear reactors.

2013   After years of fierce opposition, a never used facility for storing high level nuclear waste at a site on the Skull Valley Indian Reservation in Utah is abandoned.

2013   Experiments in inertial confinement fusion at the National Ignition Facility reach a milestone on the path to a true break-even ignition.

This glossary briefly defines scientific and technical terms involved with nuclear physics, nuclear power technology, energy use and policy, and regulation.

absorbed dose　The amount of energy deposited in a given weight of biological tissue. The units of measure are the rad and the gray.

activity　The rate at which a sample of radioactive material decays, measured in curies or becquerels.

adsorption　(not to be confused with absorption) The transfer of dissolved materials (such as radionuclide contamination in water) to a solid surface, such as a rock formation.

alpha particle　A positively charged particle made up of two protons and two neutrons that can be emitted in the radioactive decay of some heavy atoms. Because it is relatively slow, an alpha particle cannot penetrate clothing or even the outer layer of skin. However, it can be dangerous if inhaled, ingested, or emitted by some source that has entered the body.

aquifer　Geological formations that can hold or carry water easily, such as sand, gravel, or some kinds of rock. Nuclear waste stored too close to an aquifer might get into the water supply.

atom　The basic component of all chemical substances. An atom is made up of protons, neutrons, and electrons.

atomic energy　*see **nuclear energy***

atomic mass   The total number of protons and neutrons that make up the nucleus of an atom. For example, the most common form of uranium has a total of 238 protons and neutrons and thus has an atomic mass of 238.

atomic number   The number of protons or electrons in an atom, which determines what element the atom represents. For example, an atom with 92 protons and electrons has an atomic number of 92 and forms the element uranium.

backfill   In burying nuclear waste, the backfill is the soil or other material that is filled in around the waste canisters after they are placed in the ground.

background dose   The amount of radiation received by a person from natural sources. It varies with altitude and the presence of natural sources of radioactivity (such as radon).

background radiation   Radiation that either comes from space (the sun or cosmic rays) or is emitted by substances that occur naturally in the environment (such as traces of radioactive materials in construction materials and radon gas).

base loaded   A power plant maintained at its maximum capacity because it is the most economical source of power in the area.

baseline information   Social, demographic, and environmental information gathered before siting a nuclear or other energy facility.

becquerel   A unit of radioactivity equal to one atomic disintegration per second. A very tiny quantity equal to about 27 picocuries (billionths of a curie).

beta particle   A negatively charged particle that can be emitted through radioactive decay; essentially a freely moving electron. Beta radiation has short range and limited penetrating power, so it is generally dangerous only if emitted by a source close to or inside the body.

boiling water reactor (BWR)   A nuclear reactor in which water boils as it passes through the core. The boiling water serves both as a coolant (by carrying heat away from the reactor core) and as a source of steam to run a turbine to generate power. The steam is radioactive, so it must be contained and shielded within the turbine.

breeder reactor   A nuclear reactor in which the neutrons resulting from fission are used to create more new fissionable atoms than were used in the reactor. The reactor thus "breeds" its own fuel.

British thermal unit (BTU)   A unit of heat energy; the amount of heat required to raise the temperature of one pound of water by one degree Fahrenheit.

burn-up   The amount of energy produced in a power reactor per unit of original nuclear fuel. Usually expressed as megawatt-hours per metric ton.

cancer dose   The dose of radiation that if spread through a population is expected to cause one additional cancer death. The "official" estimate of the cancer dose is about 2,000 rem, but some critics believe it may be as low as 120–150 rem.

canister   The outer metal container in which glassified high-level radioactive waste or spent fuel rods are placed.

cask   A container used to transport and shield waste canisters or other radioactive materials during shipping.

chain reaction   The process by which one nuclear fission emits neutrons that in turn cause other fissions, resulting in a self-sustaining release of nuclear energy.

chemical reaction   A process in which one or more chemical substances interact, creating different substances and using or releasing energy. Unlike nuclear reactions, chemical reactions do not affect the cores of atoms and do not change one element into another.

cladding   Material (such as steel or a metal alloy) that is used to encapsulate and shield pieces of radioactive fuel in a reactor.

cold fusion   Hypothetical fusion reaction achieved by infiltrating atoms into a tightly packed crystal lattice rather than through the use of high temperature and pressure. Failure to replicate the experiments of the late 1980s has led most scientists to reject the existence of cold fusion.

colloid Small particles (10^{-9} to 10^{-6} meters in size) that are suspended in a solvent such as water. Natural colloids arise from clay minerals in contact with water, but radioactive waste can also escape into the environment in this form.

commissioning The process of making a nuclear power plant operational, which involves a lengthy testing and regulatory process.

condenser The pipes and other apparatus that allow steam to cool and turn back into water.

containment building The thick, steel-reinforced concrete building that surrounds the pressure vessel of a nuclear reactor. It represents the final barrier to the release of radioactive materials into the atmosphere.

containment system The layers of packaging and other systems designed to prevent radioactive materials from escaping into the environment during shipping.

contamination Radioactive material that has gotten into an undesired location, such as through leaks, leaching, or some other process.

control rods Long, thin rods made of a material that absorbs neutrons. Inserting rods into the reactor reduces the number of neutrons that can cause fission and thus brings the nuclear chain reaction to a halt.

control room The room in a nuclear power plant from which all operations are monitored and controlled through a large array of gauges, indicators, and controls.

conversion ratio The ratio between the fissile material created in a fission reaction and the original amount of such material. In a breeder reactor more fuel is created than is consumed, so the conversion ratio is greater than 1.

coolant A fluid (usually water) that is circulated through the core of a nuclear reactor to remove the heat generated by the fission process. Some reactors use liquid metal (sodium) or gas as a coolant.

cooling pond A pool or body of water used to dissipate heat from power plant operations through evaporation.

cooling system The system of pumps, pipes, and valves that carries coolant through a nuclear power plant.

cooling tower A tall, usually funnel-shaped, structure that removes heat from water that has condensed from steam in the condenser. Cooling the water before putting it back in the environment prevents damage to wildlife.

core The central part of a nuclear reactor where the fission chain reaction takes place. In a conventional reactor it contains fuel rods, a moderator, and control rods. A coolant is continuously circulated through the core to remove heat, which is used to generate electric power.

cosmic rays High-energy particles from space that make up much of peoples' background radiation exposure.

critical The condition in which a nuclear reaction is self-sustaining because the neutrons released from each fission can (on the average) trigger one new fission.

critical mass The minimum amount of fissionable material that for a given geometric arrangement allows for a self-sustaining nuclear reaction.

curie A unit of radioactive intensity equivalent to a grain of radium, representing 37 billion atomic disintegrations per second.

decay The disintegration of an unstable atomic nucleus, resulting in the emission of charged particles and/or photons (gamma rays).

decommission Removal from service of a nuclear power plant that has ended its service life. Includes the dismantling and safe disposal of the reactor vessel and the reduction of radioactivity at the site so that it is available for unrestricted use.

Decon   Nuclear Regulatory Commission term for decommissioning a nuclear plant by promptly decontaminating it and restoring the site to normal use.

decontamination   The process of removing radioactive materials that have become mixed with the environment, a worker's clothing, or other materials.

defense in depth   The design philosophy for nuclear facilities in which several different barriers are placed, each believed sufficient to prevent an escape of radioactive materials. For example, American reactors have both an inner containment and a strong outer building.

depleted uranium   Uranium left over from the enrichment process that segregates and concentrates uranium-235. The depleted uranium is thus much lower in uranium-235 than natural unprocessed uranium. Because of its density, depleted uranium is used to make shells or bullets that have extra penetrating power against armor.

deregulation   The process of reducing or eliminating regulation, usually in favor of increased private oversight or responsibility.

design margin   The added margin of safety built into materials or components to allow for unusual conditions and for the inherent variability in materials.

disposal   The permanent removal of radioactive materials from contact with the environment, such as by glassifying and burying them.

dose   The quantity of radiation received by an object or person, measured in rads.

dose limit   The maximum exposure to radiation permitted by regulation (e.g., for nuclear power plant workers).

dose reconstruction   The attempt to estimate the dose of radiation that a person has received through exposure to, for example, a nuclear power plant accident.

dosimeter   A device (such as a film badge) that indicates the total amount of radiation exposure received by a worker in a nuclear power plant. Government regulations specify the maximum exposure that workers are allowed to receive both in a particular amount of time and over their lifetime.

electric charge   A fundamental property of matter. Protons have a positive charge, electrons have a negative charge, and neutrons have no charge.

electric force   The force exerted by electric charge. Particles with opposite charges are attracted to one another, while those with the same charge repel each other. The attraction between protons in the nucleus and the surrounding electrons holds atoms together, but in the nucleus the nuclear force that holds protons together is stronger than the electric force that would otherwise push them apart.

electron   The smallest basic particle that makes up the atom. Electrons are negatively charged and orbit the atomic nucleus. The number and arrangement of electrons in an atom determine its chemical behavior. The flow of electrons is what makes electricity.

element   A basic substance such as hydrogen, oxygen, or uranium that cannot be chemically broken down into simpler substances. An element can be changed into another element only through a nuclear reaction.

emergency core cooling system   A safety system that is designed to prevent the core of a reactor from melting when there is a loss of coolant accident (LOCA). It usually consists of a separate emergency water source, pumps, and connecting pipes.

emission standards   Regulations that limit the amount of a substance that can be discharged into the environment.

energy   The capacity to do work. Energy can be converted from one form to another. A nuclear power plant uses the heat from nuclear reactions to create steam pressure that turns a turbine whose mechanical energy is then converted to electrical energy.

energy policy   Public policies that determine how a nation will assess and develop its energy sources.

energy security   The goal of ensuring that a nation has uninterrupted access to a sufficient supply of energy at a reasonable cost. Promoting nuclear power is seen by advocates as an important way to provide energy security by reducing reliance on imported oil.

enrichment   The process by which the proportion or concentration of a desired isotope of an element is increased. For example, most nuclear power plants use a fuel in which the proportion of uranium-235 has been enriched from its natural concentration of 1 percent to about 3 percent.

entombment   The sealing of a decommissioned or accidentally damaged nuclear power plant in concrete.

environment   The system of physical, chemical, and biological factors by which all organisms including human beings exist.

environmental impact   The effect that a given facility or its activities has on the surrounding environment, such as an increase in chemical pollution or radiation.

environmental impact statement   A legally required analysis of the impact (both short-term and long-term) that a proposed development such as a power plant will have on the surrounding environment.

exposure   Ionization produced by X-ray or gamma radiation. Exposure can be acute (intense, short-term) or chronic (low-level, of long duration).

external radiation dose   Radiation dose received from materials outside the body, mainly consisting of gamma rays.

fast neutron reactor   A reactor that uses fast, energetic neutrons. It does not require a moderator, and the extra neutrons it produces can create new fissionable atoms. However fast reactors require more highly enriched fuel that is closer to weapons-grade.

fast neutrons   High-energy neutrons such as those ejected by fission reactions.

fertile material   Material made of atoms that readily absorb neutrons and become fissionable. Fertile material in spent fuel can thus be exposed to radiation to create new fuel.

fission   The splitting of a heavy nucleus into two roughly equal parts, yielding a large amount of energy and releasing one or more neutrons that can, under the right circumstances, strike other atoms and split them in turn. Fission can occasionally occur spontaneously, but it is usually the result of a nucleus being struck by gamma rays, neutrons, or other particles.

fission products   The atoms formed by the pieces into which a heavy nucleus breaks when it fissions. These atoms are usually unstable (radioactive).

fissionable (or fissile) material   Isotopes that readily fission when struck by a neutron. Uranium-235 and plutonium-239 are the two most commonly used fissionable materials in nuclear reactors.

fossil fuels   Fuels such as coal, oil, or natural gas that are the remnants of ancient biological activity. Fossil fuels have limited supply, and all produce CO_2 that contributes to global warming.

fuel   The fissionable material that is used to create the chain reaction found in a nuclear reactor.

fuel cycle   The entire process by which fuel for nuclear reactors is obtained and used. The cycle includes mining and enriching the uranium; fabricating the fuel rods; and sometimes, reprocessing the spent fuel. The fuel cycle is complicated by the fact that reactor operation also produces new fissionable materials (such as plutonium) that can be processed into new fuel.

fuel pellet   A small cylinder of nuclear fuel (usually uranium oxide) about one-quarter inch thick and one-half inch long.

fuel rods   A rod about 12–14 feet long that contains fuel pellets. The rods are enclosed in bundles and inserted into the reactor.

fuel storage pond   A pond in which spent fuel is kept after it has been removed from a nuclear reactor. Such fuel is highly radioactive. The water serves as both a coolant and a shield against radiation getting into the environment. Once the radiation level of the fuel has declined sufficiently, it can be shipped to a permanent waste disposal facility. The failure to approve and build such facilities has resulted in a large amount of spent fuel remaining in storage ponds.

fusion   The process in which light atomic nuclei come together at high speed and combine into heavier atoms, releasing a large amount of energy. For example, two hydrogen atoms can be fused into a single helium atom.

fusion power   The use of controlled fusion reactions to create a steady supply of power. The technology needed to safely contain and maintain a fusion reaction so that it produces more power than it consumes has not yet been developed.

gamma ray   High-energy radiation with a short wavelength, similar to X-rays. Gamma rays are highly penetrating but can be stopped by a sufficient thickness of lead.

gas-cooled reactor   A reactor in which a gas such as carbon dioxide, rather than water, is used as the coolant.

gaseous diffusion   The process by which lighter uranium-235 is separated from heavier uranium-238 in uranium hexafluoride gas because the lighter atoms move more quickly through a membrane.

Geiger counter   An instrument that measures radioactivity. Radiation ionizes the gas in the tube, allowing it to conduct electricity, which is then measured and correlated to the radiation level.

generation   A broad classification of reactor designs. Early experimental reactors of the 1950s were Generation I. Most commercial reactors built in the 1970s to the 1990s were Generation II, while more recent Generation III reactors feature incremental safety and efficiency improvements. Generation IV represents designs still on the drawing board.

generation time   The time between ejection of a neutron by a fission event and its causing fission in another nucleus.

glassifying   see *vitrification*

global climate change   Changes in long-term climate believed to be brought about at least in part by the amount of carbon dioxide being released into the atmosphere by the use of fossil fuels. Such effects include a rise in average temperature, melting ice caps, a rise in sea level, and an increase in the number of violent storms. These developments may have serious effects on coastal cities, storm damage, food production, and wildlife.

global warming   Increase in average temperature around the world. Sometimes also used to mean global climate change in general (see preceding entry).

gray   A unit of absorbed radiation equal to 100 rads.

greenhouse effect   The trapping of solar radiation by gases such as carbon dioxide, leading to warming.

half-life   The amount of time required for a sample of a radioactive substance to lose half of its activity through decay. For example, it takes 24,000 years for a sample of plutonium to lose half of its radioactivity.

heavy metals   Metals that have high atomic weights (usually over 40).

heavy water   Water in which the hydrogen atoms consist of deuterium (containing a neutron as well as the usual proton). Heavy water can be used as a moderator in nuclear reactors.

high-level waste   Highly radioactive material that remains after spent nuclear fuel is reprocessed. It must be solidified and stored for a long time in a secure facility.

high-temperature gas-cooled reactor   A nuclear reactor that is cooled by helium.

highly enriched uranium (HEU)   Uranium with a high concentration of uranium-235; it is used for nuclear weapons and in some propulsion reactors in naval vessels.

hot   A colloquial term meaning highly radioactive.

ignition   in fusion, the point at which heat and pressure are sufficient to sustain the reaction without the continual addition of energy.

inertial confinement   The use of powerful laser beams to compress a fuel pellet, which in turn creates shock waves that induce nuclear fusion.

integrated waste management system   A waste management system that is designed as a single, coordinated set of equipment and procedures for the safe processing and storage of nuclear wastes.

interim storage   A temporary facility for storing nuclear wastes until a permanent storage site is available. It is often on the grounds of the nuclear plant itself.

intermediate-level waste   Waste that is too radioactive to be considered low-level waste, but does not produce significant heat from nuclear decay as does high-level waste.

ion   An atom or molecule that has gained or lost one or more electrons. Losing electrons makes a positive ion, and gaining electrons makes a negative ion.

ionization   The process by which an atom has electrons stripped away, leaving a positively charged particle.

ionizing radiation   Radiation energetic enough to ionize atoms, such as gamma or X-rays (but not visible light, radio emissions, or microwaves). Ionizing radiation can disrupt cells in organisms, cause cancer, or damage genetic material.

isotopes   Atoms of a given element that differ in atomic weight. For example, uranium has isotopes with weights of

233, 235, and 238. Isotopes can differ in their nuclear properties, such as their susceptibility to fission.

kiloton   Measure of explosive force in nuclear weapons. Equivalent to 1,000 tons of TNT.

kilowatt   One thousand watts. A measurement of the rate of generation or consumption of electricity.

kilowatt-hour   The amount of energy produced or consumed by a flow of 1,000 watts of electricity for one hour.

leaching   The dissolving of a substance through permeable materials, such as a chemical percolating through a layer of soil, or a body of water. Leaching can cause wastes or contaminants to diffuse into the environment.

liability   Legal responsibility for harm done to others. Can be insured against but has been legally limited in the case of nuclear power plants.

Light water reactor   The most common class of nuclear reactor, which uses ordinary water (as opposed to heavy water, or deuterium) as its coolant. The two most common types of light-water reactors are the boiling water reactor and the pressurized water reactor.

linear, no-threshold   The theory that risk from radiation is proportional to dose received, and that any dose, however small, introduces additional risk.

liquid metal fast breeder reactor   A nuclear breeder (fuel-producing) reactor that is cooled by the circulation of a liquid metal, usually sodium.

load   The total demand for electric power from a facility at a given time.

loss of coolant accident (LOCA)   An accident (such as a pipe rupture, blockage, or pump failure) that causes a reactor to lose coolant. This can lead to an increase in core temperature and possibly a meltdown.

low-level waste   Waste that contains only a small amount of radioactivity, such as discarded protective clothing. Such waste can usually be buried with minimal containment.

meltdown   Overheating of a reactor core (such as through loss of coolant) where it melts and sinks into the earth. This can result in a steam explosion that releases large quantities of highly radioactive material into the atmosphere.

metric ton   One thousand kilograms, or about 2,204 pounds. Often used in scientific measurements instead of ordinary "short" tons (2,000 pounds).

mill tailings   The slightly radioactive material left after uranium ore is refined to extract uranium oxide. Tailings can cause significant environmental problems such as heavy metal contamination and acidification.

millirem   A dosage of radiation equal to a thousandth of a rem. An average person receives about 100–200 millirem a year from natural and human-made radiation sources; government guidelines recommend that total exposure not exceed 500 millirem.

mixed-oxide fuel   Reactor fuel made by combining standard uranium oxide fuel with plutonium that has been obtained from dismantled nuclear weapons or extracted from spent reactor fuel.

moderator   A substance that is used to slow down the neutrons produced by fission so they are more likely to be absorbed by other atoms and perpetuate the chain reaction. Water is the most commonly used moderator; in some reactor designs the same body of water serves as both moderator and coolant. Some reactors used graphite as a moderator.

molten salt breeder reactor   A breeder reactor that uses a molten uranium-thorium salt mixture as both fuel and coolant and produces more uranium than it consumes.

monitored retrievable storage   The temporary storage of radioactive materials in such a way that it can be monitored for leaks, overheating, or other problems and can be retrieved later for shipment to a permanent storage facility. Spent reactor fuel has been kept in this form of storage in the hope that a permanent storage facility will someday be completed.

multiplication factor   The average number of neutrons from each fission reaction that go on to cause another fission. For steady operation, a reactor must have a multiplication factor of exactly one.

mutation   A genetic effect brought about by alteration of DNA, such as through exposure to natural or artificial ionizing radiation.

neutron   An atomic particle that has the same mass as a proton but has no electric charge. Protons and neutrons together make up the atomic nucleus. Because neutrons have no charge, they can approach (and possibly split) a nucleus without having to overcome electrical repulsion.

neutron poison   A material that absorbs neutrons. If such a material (such as xenon-135 or samarium-149) accumulates in a nuclear reactor as a byproduct of fission, it can halt the chain reaction.

nuclear energy   Energy released by changes in an atomic nucleus, such as by its splitting (fission) or combining (fusion).

nuclear island   The nuclear reactor complex as a whole, including containment building, auxiliary cooling system, and other apparatus.

nuclear pile   A relatively simple, early design of reactor made of bricks containing uranium and a moderating material such as graphite.

nuclear security   As defined by the International Atomic Energy Agency: the prevention and detection of, and response to, theft, sabotage, unauthorized access, illegal transfer or other malicious acts involving nuclear or other radioactive substances or their associated facilities.

nuclear steam supply system   A nuclear reactor and steam-generation system that provides steam that is used in a turbine to generate electricity.

nuclear weapon   A device that uses an arrangement of fissionable material to produce an extremely violent, uncontrolled nuclear chain reaction that creates a very powerful explosion. For greater power the initial fission explosion can be used to create a fusion (thermonuclear) reaction.

nuclear weapons capability   The ability of a nation or non-national group to design or build nuclear weapons in a relatively short period of time. Capability requires expertise, suitable equipment, and access to or the ability to create fissionable materials.

nucleon   A proton or neutron, normally found in the nucleus of an atom.

nuclide   A type of atom, specified by atomic mass, atomic number, and energy state, that has particular nuclear properties such as radioactivity intensity and half-life.

once-through fuel cycle   A simple fuel cycle in which nuclear fuel is used only once, with the spent fuel being stored as radioactive waste rather than being reprocessed to extract fissionable uranium-235 or plutonium-239.

operating costs   The cost of routine operation of a nuclear power plant.

operating license   The legal document that permits a nuclear power plant to be operated for commercial power generation.

pathway analysis   Analysis of the routes or processes by which radioactive or toxic material can reach human beings.

peak load   The maximum level of demand for electricity from a power plant measured over a period of time such as a day, season, or year.

percolation   The downward movement of water (and whatever is dissolved in it) through permeable layers of rock.

photon   The fundamental unit (quantum) of electromagnetic radiation, having a characteristic energy level ranging from radio to light to X-rays and gamma rays.

plutonium   An element that is artificially produced by bombarding uranium with neutrons in the core of a reactor. Plutonium in turn is able to fission easily and can be used as reactor fuel or for nuclear weapons.

pollution   In general, the deliberate or accidental introduction of harmful substances, heat, or radiation into the environment (ground, water, or atmosphere).

power plant   Any facility that generates electricity from some form of energy. A nuclear power plant uses the heat from splitting atoms (fission) to generate steam that turns a turbine to generate electricity.

pressure vessel   The steel enclosure that surrounds the core of a reactor. It is designed to withstand high pressures and temperatures to prevent radioactive material from escaping in the event of an accident involving the core.

pressure tube reactor   A nuclear reactor in which fuel is placed inside a large number of tubes through which is circulated a high-pressure coolant.

pressurized water reactor   A nuclear reactor that is cooled by water that is kept under high pressure to prevent it from boiling, allowing it to carry a large amount of heat. The coolant then heats the water in a separate loop from which steam for the turbine is produced.

primary coolant   The liquid or gas that provides most of the cooling for the fuel elements in the reactor core.

primary loop   The part of a reactor's cooling system that sends coolant through the reactor core.

proliferation   The spread of nuclear weapons, nuclear weapons capability, or weapons components (including fissionable material) to nations not currently having nuclear capability.

proton   A positively charged particle that together with the neutron makes up the nucleus of the atom.

public policy   Plans and procedures established by a government agency to deal with issues of public concern, such as the development of nuclear power, the storage of nuclear waste, and the prevention of nuclear proliferation.

rad (radiation absorbed dose)   A measure of the amount of ionizing radiation that has been absorbed by any material, including human tissue.

radiation   Particles or electromagnetic waves that go out from a source into surrounding space. The disintegration of an unstable atom produces radiation.

radiation hormesis   The hypothesis that low levels of ionizing radiation benefit organisms by stimulating the cellular repair mechanism and helping to protect against disease.

radiation sickness   Illness caused by exposure to a large dose of radiation (100 rem or more). Symptoms include nausea, vomiting, intestinal bleeding, and anemia.

radiation therapy   Use of radiation in medicine, such as to kill some types of tumors.

radioactive tracer   Small amount of radioactive material that can be used to track the path of a substance through the body.

radioactive waste   Unwanted radioactive materials produced by a nuclear facility, or ordinary materials that have been mixed with radioactive substances or made radioactive through exposure to gamma rays or other radiation.

radioactivity   The spontaneous emission of alpha or beta particles (or gamma rays) by atoms of unstable elements, such as uranium.

radioisotope   An unstable isotope of an element, which will eventually undergo radioactive decay.

radionuclide   Any radioactive isotope.

radon   A radioactive gas that is naturally produced through decay of a radium decay product in the earth. It can accumulate in unventilated building spaces (such as basements). Prolonged breathing of radon can induce lung cancer. Radon represents more than half of the average background radiation dose.

reactor   A device that contains a controlled nuclear fission chain reaction. The reaction can be used to generate electricity, propel a ship, carry on nuclear research, or produce plutonium for nuclear weapons.

reactor vessel   The container of a reactor's nuclear core. It can be a pressure vessel, a low-pressure vessel, or a prestressed concrete enclosure.

reactor year   The operation of a nuclear reactor for one year. Often used in statistics such as accidents per reactor year or expected reactor years between major accidents.

recycling   The extraction and reuse of remaining fissionable material in spent fuel.

refueling   The removal of spent fuel from a reactor and its replacement with new fuel.

regulatory policy   Policies that specify each aspect of a nuclear plant's operations, including training requirements, safety standards, reporting requirements, and releases into the environment.

rem   Acronym for "roentgen equivalent man," a unit of radiation effect on the human body.

renewable energy source   Sources of energy that last indefinitely without being depleted, such as solar and hydroelectric power. Uranium is not fully renewable as an energy source, but its life can be extended through use of breeder reactors.

repair enzymes   Substances in the body that may be able to repair damage to DNA caused by low doses of radiation.

repository   A permanent storage facility for nuclear wastes, including spent fuel.

reprocessing   The chemical treatment of spent nuclear fuel to remove remaining uranium and plutonium from the fission products, allowing it to be used to make new fuel.

roentgen   The unit of exposure to gamma rays or X-rays, named for Wilhelm Roentgen, who discovered X-rays in 1895.

safeguards   Regulations and procedures for monitoring and controlling the storage and shipping of nuclear materials to prevent their being diverted by unauthorized persons or groups, including terrorists and nations seeking nuclear weapons.

safety system   In general, a combination of electronic, mechanical, and instrumentation systems designed to detect or prevent accidents that might endanger the reactor or the surrounding community.

Safstor   Nuclear Regulatory Commission term for decommissioning a nuclear plant by "mothballing" it, that is, letting it run down gradually until has lost most of its radioactivity.

scram   A rapid shutdown of a reactor core by inserting neutron absorbers (usually rods) when there is danger of the reaction getting out of control. Since this can happen very quickly, it is usually an automatic process triggered by specified conditions rather than being triggered by the operator.

secondary loop   The part of a nuclear power plant that is not in contact with the core, but uses heat carried from the core to generate steam to produce electricity.

shielding   Materials such as water, lead, or concrete that are placed around radioactive material to protect personnel and the environment from radiation exposure.

slow neutrons   Neutrons that are slower than those ejected by a fission event. Slow neutrons are much more effective than fast neutrons for causing fission in uranium-235, so reactors use a moderator to slow down the neutrons produced by fission.

small modular reactor   One of a number of designs for small, self-contained, relatively low power reactors that can be manufactured and then moved to the site where they are to be operated.

spent fuel   Nuclear fuel whose fissionable material has been depleted to the point where it can no longer sufficiently sustain a chain reaction. Spent fuel is both thermally hot and highly radioactive and must be handled with extreme caution.

spent fuel pool   A deep pool of water near a nuclear reactor in which spent fuel can be stored until it cools enough to allow it to be moved to a more permanent storage facility.

stable nucleus   A nucleus that can exist indefinitely without spontaneously coming apart; it is thus nonradioactive.

steam generator   A device that uses the heat from pressurized water or hot gas to create steam for use in power generation.

storage   The process of isolating, containing, and potentially recovering radioactive materials. Storage facilities must be designed so that they can be monitored and maintained for a specific period of time.

thermal pollution   The discharge of excess heat, such as into rivers or lakes. This can harm living things and disrupt the ecosystem.

thermal neutron   A type of slow neutron that is in thermal equilibrium with surrounding material (such as a moderator).

thermal reactor   A nuclear reactor where the chain reaction is sustained mainly by thermal neutrons.

thorium-232   A natural isotope that can be bombarded to create the fissionable isotope uranium-233.

threshold theory   The idea that there is some level of total radiation exposure below which there are no health effects. Most scientists now believe there is no such threshold, although effects of low doses of radiation are hard to prove.

tokamak   A donut-shaped magnetic device used to contain a nuclear fusion reaction.

transuranic wastes   Nuclear wastes that consist of heavy elements that are derived from uranium, isotopes of plutonium, and other elements with atomic numbers greater than 92 (uranium). Such wastes are produced mainly from reprocessing spent fuel and form use of plutonium in building nuclear weapons.

turbine   An engine that consists of blades attached to a rotating shaft. It can be used to convert the kinetic energy from steam to mechanical motion that in turn can be used to generate electricity.

uranium   A natural radioactive element that is heavy, hard, shiny, and metallic. Certain isotopes of uranium are suitable for use in nuclear reactors.

uranium-233   A fissionable uranium isotope suitable for use in making nuclear weapons.

uranium-235   A fissionable uranium isotope that makes up less than 1 percent of all natural uranium. It is the most common fuel used in nuclear power plants.

uranium-238   The most common (99 percent) isotope of uranium. It is not very fissionable but can be turned into the fissionable isotope plutonium-239 in a nuclear reactor.

vitrification   A process that turns radioactive wastes into a glasslike substance that is relatively inert and easy to store.

waste disposal system   The overall system of processing facilities, storage sites, and transportation systems for handling radioactive wastes.

water table   The upper boundary of the area beneath the ground surface that is completely saturated with water.

watt   A unit of power (the rate at which work is done). For electricity, it is a measure of the rate of generation or consumption. Often used with the prefixes "kilo-" (thousand), "mega-" (million), or "giga-" (billion).

watt-hour   The amount of energy involved in generating or consuming one watt of electricity for one hour.

weapons-grade uranium   Highly enriched uranium (from 20 to 70 percent uranium-235) that can be used to make nuclear weapons. (Commercial enriched uranium for power plants is only about 3.5 percent uranium-235).

yellowcake   A uranium-rich yellow powder that is the residue left after pouring crushed uranium ore into an acid that dissolves the uranium, allowing the acid solution to evaporate.

363

About the Author

Harry Henderson has been a science and technical writer for more than 30 years. His specialties include computing, mathematics, physical science, and the social impact of technology. Recent publications include biographies of Richard Feynman and Alan Turing, as well as the *Encyclopedia of Computer Science and Technology*, Second Edition. (Facts on File, 2008)